CliffsQuickReview

Biochemistry II

By Frank Schmidt, Ph.D.

Wiley Publishing, Inc.

About the Author

Frank Schmidt is a professor of Biochemistry at the University of Missouri-Columbia.

Publisher's Acknowledgments

Editorial

Project Editor: Sherri Fugit

Acquisitions Editor: Kris Fulkerson

Copy Editor: Corey Dalton

Technical Editor: Jessica Joy Hollenbeck

Editorial Assistant: Laura Jefferson

Composition

Wiley Indianapolis Composition Services

CliffsQuickReview™ Biochemistry II

Published by:
Wiley Publishing, Inc.
111 River Street
Hoboken, NJ 07030
www.wiley.com

Library of Congress Control Number: 00-103372

ISBN: 978-0-7645-8562-3

10 9 8 7 6 5 4

1O/RV/QS/QX/IN

Published by Wiley Publishing, Inc., New York, NY
Published simultaneously in Canada

For general information on our other products and services or to obtain technical support, please contact our Customer Care Department within the U.S. at 800-762-2974, outside the U.S. at 317-572-3993, or fax 317-572-4002.

Wiley also publishes its books in a variety of electronic formats. Some content that appears in print may not be available in electronic books.

CONTENTS

CONTENTS

CONTENTS

CONTENTS

FUNDAMENTAL IDEAS

For the purpose of your review, your knowledge of the following fundamental ideas is assumed:

- The scope of biochemistry
- Biological energy flow
- Protein structure
- Weak interactions
- Physiological chemistry of oxygen binding
- Enzymes
- Organization of metabolism
- Glycolysis
- Tricarboxylic acid cycle
- Oxidative phosphorylation
- Carbohydrate metabolism II

If you need to review any of these topics, refer to *CliffsQuickReview Biochemistry I.*

CHAPTER 1
FATTY ACID OXIDATION

Fats and Oils

Triacylglycerols (fats and oils) store the majority of the energy in most animals and plants. Fats such as beef tallow remain solid or semisolid at room temperature while oils such as olive oil or corn oil are liquid at that temperature. Oils solidify only at lower temperatures—in a refrigerator, for example. The different kinds of fatty acids found in the side chains of the triacylglycerol cause the different melting temperatures. The fatty acids of oils contain more double bonds than do those of fats.

These molecules make good energy-storing units because their oxidation releases more energy than the oxidation of carbohydrates or amino acids. The **caloric density** of triacylglyerols is about 9 kilocalories per gram, compared to 5 kilocalories per gram for the latter biomolecules.

Essential fatty acids are precursors to membrane lipids and to compounds that serve as intercellular signals in animals.

The caloric density of lipids is due to the side chain carbons of fats being more reduced (hydrogen-rich) than the side chain carbons of carbohydrates, for example:

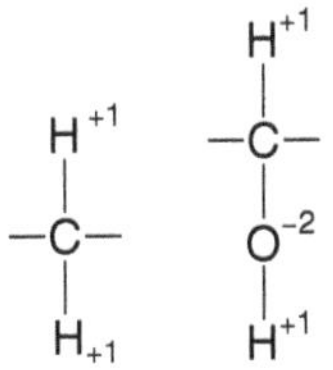

H^{+1} | —C— | H_{+1} H^{+1} | —C— | O^{-2} | H^{+1}

Fat, oxidation state = –2 → Carbohydrate, oxidation state = 0

Oxidation of the carbon found in fatty acids to carbon dioxide involves a change in oxidation number from –2 to +4, while the oxidation of the carbon of carbohydrate involves a change from 0 to +4. The greater change in oxidation number means that the oxidation of fat releases more energy. (This is a general principle; for example, burning methane, CH_4, releases more heat than burning methanol, CH_2OH). On the other hand, while amino acids and carbohydrates can oxidize anaerobically (without added oxygen), fats can oxidize only aerobically. Many cultures have used this characteristic to preserve foods in animal fat. The fat prevents the growth of oxygen-requiring molds and bacteria.

Dietary Fat Absorption

Energy production from triacylglycerols starts with their hydrolysis into free fatty acids and glycerol. Enzymes called **lipases**, which catalyze the reaction, carry out this hydrolysis.

$$
\begin{array}{l}
\text{H}_2\text{C—O—C(=O)—R}_1 \\
\text{H—C—O—C(=O)—R}_2 \\
\text{H}_2\text{C—O—C(=O)—R}_3
\end{array}
\;+\; 3H_2O \longrightarrow
\begin{array}{l}
\text{H}_2\text{C—OH} \\
\text{HC—OH} \\
\text{H}_2\text{C—OH}
\end{array}
\;+\;
\begin{array}{l}
\text{H—O—C(=O)—R}_1 \\
\text{HO—C(=O)—R}_2 \\
\text{HO—C(=O)—R}_3
\end{array}
$$

The reaction releases the three fatty acids and glycerol. An intestinal carrier absorbs the glycerol, which will eventually rejoin with fatty acids in the intestinal cells.

The body must absorb the fatty acids released by the lipases by a rather more involved mechanism. Fatty acids are poorly soluble in water, although they are more soluble than triacylglycerols. Lipids of whatever kind tend to form droplets. Protein enzymes are water-soluble and therefore cannot gain easy access to the lipid droplet. To be digested, lipids must be *emulsified* into small droplets, which have a larger surface area. In other words, the hydrophobic interactions forcing the lipids into larger droplets must be overcome. The molecules that carry out this function are called **bile salts** or **bile acids**. Metabolically, the liver creates them and secretes them into the gall bladder, from where they are pumped into the duodenum.

Bile salts are derived from cholesterol and are a major end product of cholesterol metabolism. They are powerful detergents, with a large, hydrophobic component and a carboxylic acid end-group that is negatively charged at the pH characteristic of the small intestine. The hydrophobic component of the bile acid will associate above a specific concentration (termed the **critical micelle concentration**, or **CMC**) to form disc-shaped **micelles**, that is, droplets. Common bile salts have CMCs in the 2 to 5 (millimolar) range. The micelles in the gut contain dietary lipids (triacylglycerol, cholesterol, and fatty acids) as well as bile salts, and are termed **mixed micelles** for that reason. Figure 1-3 shows a diagram of a mixed micelle.

Figure 1-1

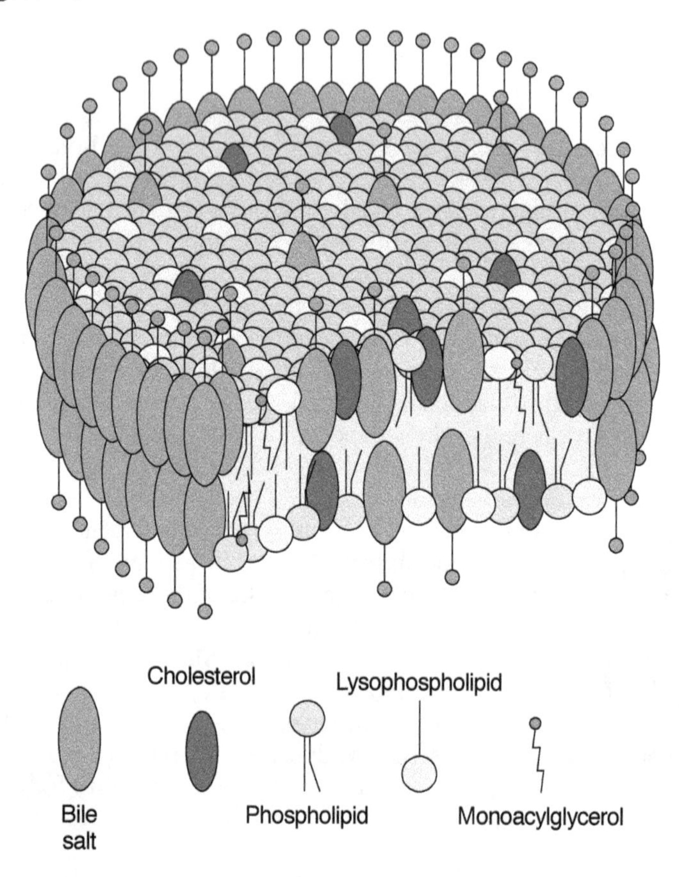

The bile salts form the edge of the micelle and also appear, in fewer numbers, dispersed throughout the inside of the micelle. The lipids exist in a bilayer on the inside of the disc. Bile acids are important for fatty acid absorption. Fat-soluble vitamins (A, D, E, and K) absolutely require bile acids for absorption.

The mixed micelle provides a large surface area for the action of **pancreatic lipase**, which is responsible for the majority of digestive action. Pancreatic lipase uses a cofactor, a small protein called **colipase**, which binds both to lipase and to the micelle surface. The action of lipase leads to free fatty acids that are slightly soluble in the aqueous phase of the gut. For the most part, the cells of the small intestine absorb these free fatty acids; the bacteria in the large intestine metabolize and/or absorb those that pass through the small intestine. The bile salts are reabsorbed in the last third of the small intestine.

Bile-acid metabolism explains the ability of certain kinds of dietary fiber to help lower serum cholesterol. A molecule of bile acid circulates through the liver and intestine five or more times before finally being eliminated. Soluble fiber (such as that found in oat bran) binds bile acids, but itself cannot be absorbed. Therefore, fiber-bound bile acids are eliminated in the stool. Because bile acids derive from cholesterol, synthesizing more bile acid drains the body's stores of cholesterol, which leads to a reduction in serum cholesterol, and therefore, to a lower risk of coronary artery disease. Eating oat fiber cannot overcome an excessive dietary cholesterol consumption, of course. In other words, consuming excessive amounts of well-marbled steak and expecting to overcome the effects by eating a bran muffin would be foolish.

Lipids in the bloodstream

Free fatty acids are transported as complexes with serum albumin. Cholesterol, triacylglycerols, and phospholipids are transported as protein-lipid complexes called **lipoproteins**. Lipoproteins are spherical, with varying amounts and kinds of proteins at their surfaces. The protein components, of which at least ten exist, are called **apolipoproteins**. Lipoproteins are classified in terms of their density.

The lightest and largest of the apolipoproteins are the **chylomicrons**, which are less dense than water by virtue of their being composed of more than 95 percent lipid by weight (remember that oils float on water because they are less dense than water).

Triacylglycerols make up most of the lipid component of chylomicrons, with small amounts of phospholipid and cholesterol. Chylomicrons contain several kinds of apolipoproteins.

Very-Low-Density Lipoproteins (VLDL) are less dense than chylomicrons. They contain more protein, although lipids (fatty acids, cholesterol and phospholipid, in that order) still make up 90 to 95 percent of their weight. Low-density lipoproteins (LDLs) are about 85 percent lipid by weight and contain more cholesterol than any other kind of lipid. VLDL and LDL contain large amounts of Apolipoprotein B. The VLDL and LDL are sometimes referred to as "bad cholesterol" because elevated serum concentrations of these lipoproteins correspond with a high incidence of artery disease (stroke and heart disease). The LDLs carry cholesterol and fatty acids to sites of cellular membrane synthesis.

High-density lipoproteins (HDLs) contain a different apolipoprotein form, Apolipoprotein A. These proteins are about half lipid and half protein by weight. Phospholipids and cholesterol esters are the most important lipid components. HDL is sometimes referred to as "good cholesterol" because a higher ratio of HDL to LDL corresponds to a lower rate of coronary artery disease.

In summary, triacylglycerols from the diet are digested by lipase and associate with bile salts into mixed micelles. The free fatty acids are absorbed by the cells of the small intestine, from which they are transported via the lymph system to the liver. From the liver, they are released as apolipoproteins in the circulation, carrying fatty acids and cholesterol to the cells throughout the body.

Tissues and lipids

The triacylglycerols in chylomicrons and LDLs circulate through the blood system; the former carries dietary lipids while the latter carries lipids synthesized by the liver. The triacylglycerols are substrates for cellular lipases, which hydrolyze them to fatty acids and glycerol in several steps. Carrier proteins transport the lipids into the cell. Different carriers exist for different chain-length lipids.

Hydrolysis of Triacylglyerols

Energy production from triacylglycerols starts with their hydrolysis into free fatty acids and glycerol. In **adipose** (fat-storing) tissue, this hydrolysis is carried out by a cellular lipase, which catalyzes the hydrolysis reaction to release the free fatty acids and glycerol. The fatty acid is carried through the bloodstream by being adsorbed to serum albumin, while the glycerol goes to the liver. In the liver, glycerol can be sent to the glycolytic pathway by the action of two enzymes, glycerol kinase and glycerol-3-phosphate dehydrogenase. Glyceraldehyde-3-phosphate can also be used as a source of glucose or, after conversion to phosphoenolpyruvate, as a source of tricarboxylic acid cycle (TCA-cycle) intermediates (see Chapter 12, Volume 1).

βeta-Oxidation

In target tissues, fatty acids are broken down through the **β-oxidation pathway** that releases 2-carbon units in succession. For example, palmitic acid has 16 carbons. Its initial oxidation produces eight acetyl-Coenzyme A (CoA) molecules, eight reduced FAD molecules, and eight NADH molecules. The fatty acid is first **activated** at the outer mitochondrial surface by conjugating it with CoA, then

transported through the inner mitochondrial membrane to the matrix, and then, for each 2-carbon unit, broken down by successive *dehydrogenation, water addition, dehydrogenation,* and *hydrolysis* reactions.

Activation: Fatty Acid $\rightleftharpoons$ Fatty Acyl-CoA

The first step in utilizing the fatty acid molecule for energy production is the conversion of the fatty acid to a CoA molecule in a two-step process:

$$R-\overset{O}{\overset{\|}{C}}-OH + ATP + CoASH \longrightarrow R-\overset{O}{\overset{\|}{C}}-SCoA + AMP + PP_i$$

Note that the hydrolysis of two high-energy phosphate bonds in ATP provides the energy source for the reaction. The inorganic pyrophosphate, PP_i, is subsequently broken down to two phosphate ions by inorganic pyrophosphatase. The action of this enzyme means that very little PP_i remains in the cell, making the synthesis of the fatty acyl-CoA favored. This is an example of **metabolic coupling**, the process whereby a thermodynamically unfavored reaction is allowed because it shares an intermediate (in this case PP_i) with a favored one.

Transport: The Role of Carnitine

While short-chain fatty acids can move across the mitochondrial membrane directly and are then activated in the mitochondrial matrix, the inner mitochondrial membrane is impermeable to longer fatty acids, such as palmitate. A small molecule, **carnitine**, serves as a carrier across the mitochondrial membrane. This pathway requires no chemical energy supply; rather, the fact that the fatty acid in the mitochondrial matrix is being broken down by oxidation drives the process.

Carnitine acyltransferase I, which is located on the outer mitochondrial membrane, transfers the fatty acyl group from fatty acyl-CoA to the hydroxyl (OH) group of carnitine. The acyl-carnitine then moves across the intermembrane space to a translocase enzyme, which, in turn, moves the acyl-carnitine to carnitine acyltransferase II, which exchanges the carnitine for Coenzyme A.

Figure 1-2

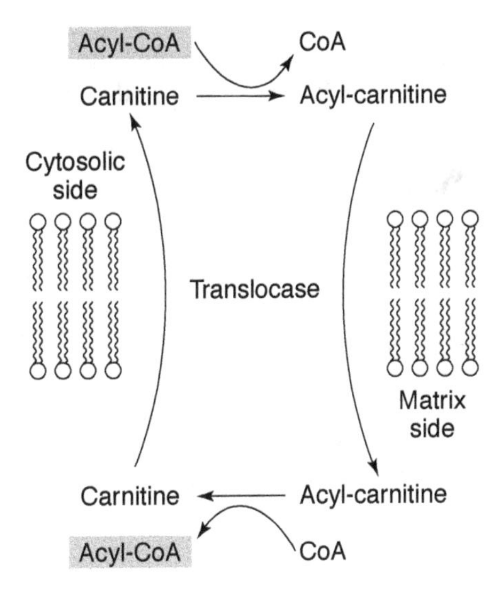

Carnitine is used as a dietary supplement by endurance athletes and in the treatment of certain metabolic diseases. Apparently the extra carnitine allows more rapid transport of fatty acids and a more efficient energy metabolism.

The β-Oxidation Helical Scheme for Fatty Acyl-CoA

In the mitochondrion, even-numbered fatty acyl-CoA is broken into acetyl-CoA units starting from the carboxyl end. The first reaction is **dehydrogenation** by an FAD-dependent dehydrogenase to form an enoyl-CoA.

Reaction 1:

$$R{-}CH_2{-}CH_2{-}CH_2{-}\overset{O}{\overset{\|}{C}}{-}SCoA \xrightarrow{FAD \;\to\; FADH_2} R{-}CH_2{-}CH{=}CH{-}\overset{O}{\overset{\|}{C}}{-}SCoA$$

This reaction absolutely depends on the fatty acids being activated by CoA. This reaction is very similar to the succinate dehydrogenase step of the Krebs cycle.

Succinate dehydrogenase:

$$HO{-}\overset{O}{\overset{\|}{C}}{-}CH_2CH_2{-}\overset{O}{\overset{\|}{C}}OH \xrightarrow{FAD \;\to\; FADH_2} HO{-}\overset{O}{\overset{\|}{C}}{-}CH{=}CH{-}\overset{O}{\overset{\|}{C}}{-}OH$$

The enoyl-CoA is then a substrate for the addition of water across the carbon-carbon double bond. This results in a β-hydroxy-acyl-CoA compound, because the OH of water is added to the carbon further away from the carboxyl group:

Reaction 2:

$$R{-}CH_2{-}CH{=}CH{-}\overset{O}{\overset{\|}{C}}{-}SCoA \xrightarrow{H_2O} R{-}CH_2{-}CH(OH){-}CH_2{-}\overset{O}{\overset{\|}{C}}{-}SCoA$$

Again, this type of reaction occurs in the Krebs cycle, with the addition of water to fumarate to make malate.

fumarase:

$$HO-\overset{O}{\overset{\|}{C}}-\underset{H}{C}=\overset{H}{C}-\overset{O}{\overset{\|}{C}}-OH \xrightarrow{H_2O} HO-\overset{O}{\overset{\|}{C}}-\underset{H}{\overset{H}{C}}-\underset{H}{\overset{OH}{C}}-\overset{O}{\overset{\|}{C}}-OH$$

The hydrogens of the β-hydroxy group are removed in a *dehydrogenation* reaction, this time using NAD as the electron acceptor.

Reaction 3:

$$R-\underset{H}{\overset{H}{C}}-\underset{H}{\overset{OH}{C}}-\underset{H}{\overset{H}{C}}-\overset{O}{\overset{\|}{C}}-SCoA \xrightarrow{NAD \rightarrow NADH_2} R-\underset{H}{\overset{H}{C}}-\overset{O}{\overset{\|}{C}}-\underset{H}{\overset{H}{C}}-\overset{O}{\overset{\|}{C}}-SCoA$$

This also occurs in the Krebs cycle, as in the dehydrogenation of malate to oxaloacetate.

malate dehydrogenase:

$$HO-\overset{O}{\overset{\|}{C}}-CH_2-\underset{H}{\overset{OH}{C}}-\overset{O}{\overset{\|}{C}}-OH \xrightarrow{NAD \rightarrow NADH_2} HO-\overset{O}{\overset{\|}{C}}-CH_2-\overset{O}{\overset{\|}{C}}-\overset{O}{\overset{\|}{C}}-OH$$

The final step in the removal of two carbons from the fatty acid is the **thiolytic cleavage** to release acetyl-CoA. The term "thiolytic" refers to the use of Coenzyme A to bond with the carbonyl carbon of the β-keto acid.

Reaction 4:

$$R{-}\overset{H}{\underset{H}{C}}{-}\overset{O}{\overset{\|}{C}}{-}\overset{H}{\underset{H}{C}}{-}\overset{O}{\overset{\|}{C}}{-}SCoA + CoASH \longrightarrow R{-}\overset{H}{\underset{H}{C}}{-}\overset{O}{\overset{\|}{C}}{-}SCoA + H_3C{-}\overset{O}{\overset{\|}{C}}{-}SCoA$$

This step leaves two cleavage products. The first, derived from the two carbons at the carboxyl end of the fatty acid, is acetyl-CoA, which can be further metabolized in the TCA cycle. The second cleavage product is a shorter fatty acyl-CoA. Thus, for example, the initial step of digesting a fatty acid with 16 carbons is an acyl-CoA molecule where the acyl group has 14 carbons and a molecule of acetyl-CoA. The β-oxidation scheme may be used to accommodate unsaturated fatty acids also. The reactions occur as described previously for the saturated portions of the molecule. Where a **trans** carbon-carbon double bond occurs between the χ- and β-carbons of the acyl-CoA, the accommodation is fairly simple: reaction 1 isn't needed. Where the double bonds are in the **cis** configuration, or are between the β and γ carbons, **isomerase** enzymes change the location of the double bonds to make recognizable substrates for β-oxidation.

Acetyl-CoA from fatty acid oxidation enters the TCA cycle in the same way as does acetyl-CoA derived from glucose: addition to oxaloacetate to make citrate. This can cause complications if an individual is metabolizing only fat, because the efficient metabolism of fat requires a supply of TCA-cycle intermediates, especially dicarboxylic acids, which can't (usually) be made from fatty acids. These intermediates must be supplied by the metabolism of carbohydrates, or more often, amino acids derived from muscle tissue.

Energy Yield from Fatty Acid Oxidation

What is the ***total*** energy yield (number of ATPs) from oxidizing one molecule of palmitic acid to CO_2? Palmitic acid has 16 carbons, so you can break it down into eight acetyl-CoA molecules, with the formation of one $FADH_2$ and one NADH at each of the seven β-oxidation steps. Electron transport starting with $FADH_2$ yields two ATPs and with NADH yields three ATPs. Therefore, β-oxidation yields the equivalent of 35 ATPs per molecule of palmitic acid. Each acetyl-CoA goes into the TCA cycle, where its metabolism yields three NADHs, one FAD, and one GTP directly, for a total of 12 ATPs. Thus, the ATPs produced are:

$$7 \times 5 + 8 \times 12 = 35 + 96 = 131$$

Two ATP equivalents were used to activate the fatty acid, leading to a total energy yield of 129 ATPs, over three times the amount of energy obtained from metabolizing a single molecule of glucose.

Odd-Numbered Chain and Branched Fatty Acids

Although the fatty acid oxidation scheme works neatly for even-numbered chain lengths, it can't work completely for fatty acids that contain an odd number of carbons. β-oxidation of these compounds leads to *propionyl-CoA* and acetyl-CoA, rather than to two acetyl-CoA at the final step. The propionyl-CoA is not a substrate for the TCA cycle or other simple pathways. Propionyl-CoA undergoes a *carboxylation* reaction to form *methylmalonyl-CoA*. This reaction requires biotin as a cofactor, and is similar to an essential step in fatty acid biosynthesis. Methylmalonyl-CoA is then isomerized by an epimerase and then by *methylmalonyl-CoA mutase*—an enzyme that uses Vitamin B_{12} as a cofactor—to form succinyl-CoA, which is a TCA-cycle intermediate.

Figure 1-3

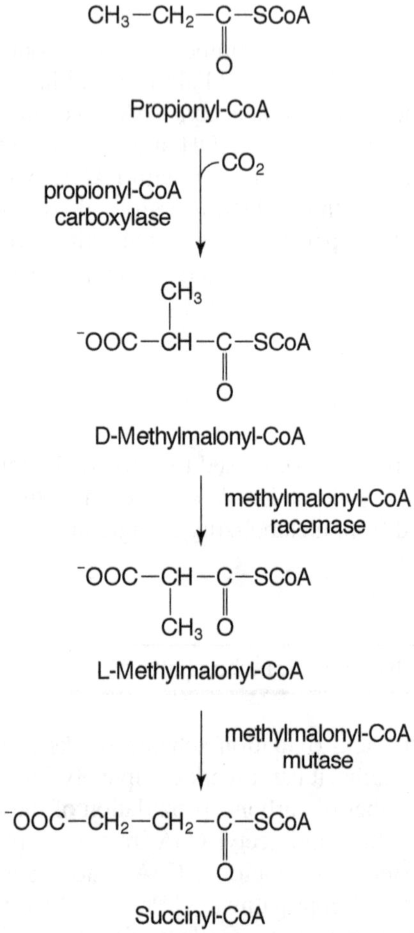

$CH_3-CH_2-C(=O)-SCoA$

Propionyl-CoA

CO_2

propionyl-CoA
carboxylase

$^-OOC-CH(CH_3)-C(=O)-SCoA$

D-Methylmalonyl-CoA

methylmalonyl-CoA
racemase

$^-OOC-CH(CH_3)-C(=O)-SCoA$

L-Methylmalonyl-CoA

methylmalonyl-CoA
mutase

$^-OOC-CH_2-CH_2-C(=O)-SCoA$

Succinyl-CoA

Branched-chain fatty acids present a problem of a similar kind. For example, phytanic acid, found in animal milk, can't be oxidized directly by β-oxidation because the addition of water is a problem at the branched β-carbon.

The first step in the digestion of this compound is the oxidation of the χ **carbon** by molecular oxygen. Then the original carboxyl group is removed as CO_2, leaving a shorter chain. This chain can now be accommodated by the β-oxidation reactions, because the new β-carbon now lacks a methyl group. (Note that the branch points yield propionyl-CoA, and must enter the TCA cycle through methylmalonate.

Figure 1-4

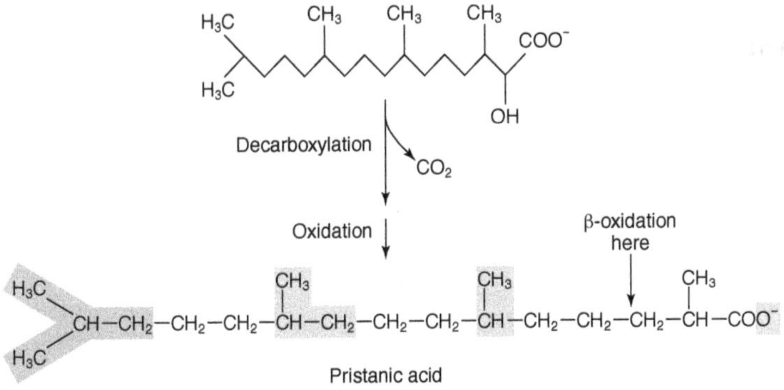

H3C
CH3 CH3 CH3
COO⁻
H3C
OH
Decarboxylation
CO2
Oxidation
β-oxidation here
H3C CH3 CH3 CH3
CH–CH2–CH2–CH2–CH–CH2–CH2–CH2–CH–CH2–CH2–CH2–CH–COO⁻
H3C
Pristanic acid

Ketone Bodies

The term "ketone bodies" refers primarily to two compounds: acetoacetate and β-hydroxy-butyrate, which are formed from acetyl-CoA when the supply of TCA-cycle intermediates is low, such as in periods of prolonged fasting. They can substitute for glucose in skeletal muscle, and, to some extent, in the brain. The first step in ketone body formation is the condensation of two molecules of acetyl-CoA in a reverse of the thiolase reaction.

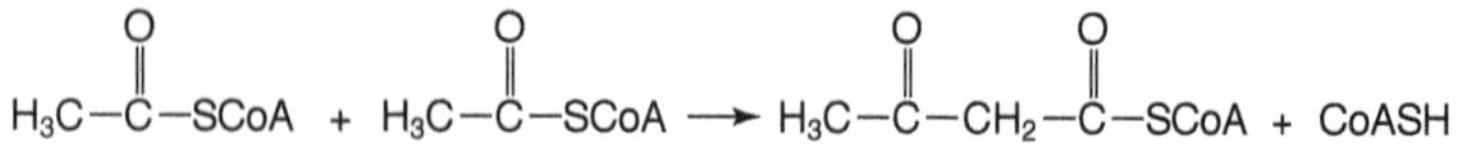

$$H_3C-\overset{O}{\overset{\|}{C}}-SCoA + H_3C-\overset{O}{\overset{\|}{C}}-SCoA \longrightarrow H_3C-\overset{O}{\overset{\|}{C}}-CH_2-\overset{O}{\overset{\|}{C}}-SCoA + CoASH$$

The product, acetoacetyl-CoA, accepts another acetyl group from acetyl-CoA to form β-hydroxy-β-hydroxymethylglutaryl-CoA (HMG-CoA). HMG-CoA has several purposes: It serves as the initial compound for cholesterol synthesis or it can be cleaved to acetoacetate and acetyl-CoA. Acetoacetate can be reduced to β-hydroxybutyrate or can be exported directly to the bloodstream. Acetoacetate and β-hydroxybutyrate circulate in the blood to provide energy to the tissues.

Acetoacetate can also spontaneously decarboxylate to form acetone:

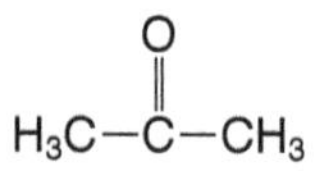

$$H_3C-\overset{O}{\overset{\|}{C}}-CH_3$$

Although acetone is a very minor product of normal metabolism, diabetics whose disease is not well-managed often have high levels of ketone bodies in their circulation. The acetone that is formed from decarboxylation of acetoacetate is excreted through the lungs, causing characteristic "acetone breath."

CHAPTER 2
LIPID BIOSYNTHESIS

Energy Storage

Fatty acid synthesis is regulated, both in plants and animals. Excess carbohydrate and protein in the diet are converted into fat. Only a relatively small amount of energy is stored in animals as glycogen or other carbohydrates, and the level of glycogen is closely regulated. Protein storage doesn't take place in animals. Except for the small amount that circulates in the cells, amino acids exist in the body only in muscle or other protein-containing tissues. If the animal or human needs specific amino acids, they must either be synthesized or obtained from the breakdown of muscle protein. Adipose tissue serves as the major storage area for fats in animals. A normal human weighing 70 kg contains about 160 kcal of usable energy. Less than 1 kcal exists as glycogen, about 24 kcal exist as amino acids in muscle, and the balance—more than 80 percent of the total—exists as fat.

Plants make oils for energy storage in seeds. Because plants must synthesize all their cellular components from simple inorganic compounds, plants—but usually not animals—can use fatty acids from these oils to make carbohydrates and amino acids for later growth after germination.

Fatty Acid Biosynthesis

The biosynthetic reaction pathway to a compound is usually not a simple opposite of its breakdown. Chapter 12 of Volume 1 discusses this concept in regard to carbohydrate metabolism and gluconeogenesis. In fatty acid synthesis, acetyl-CoA is the direct precursor only of the methyl end of the growing fatty acid chain. All the other carbons come from the acetyl group of acetyl-CoA but only after it is modified to provide the actual substrate for fatty acid synthase, malonyl-CoA.

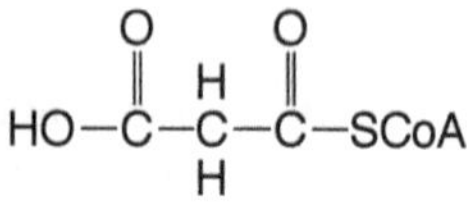

$$HO-\overset{\overset{O}{\|}}{C}-\underset{H}{\overset{H}{C}}-\overset{\overset{O}{\|}}{C}-SCoA$$

Malonyl-CoA contains a 3-carbon dicarboxylic acid, malonate, bound to Coenzyme A. Malonate is formed from acetyl-CoA by the addition of CO_2 using the biotin cofactor of the enzyme acetyl-CoA carboxylase.

$$HCO_3^- \text{ Acetyl-CoA} + HCO_3^- + \text{ATP} \rightleftharpoons \text{Malonyl-CoA} + \text{ADP} + P_i$$

Formation of malonyl-CoA is the **commitment step** for fatty acid synthesis, because malonyl-CoA has no metabolic role other than serving as a precursor to fatty acids.

Fatty acid synthase (FAS) carries out the chain elongation steps of fatty acid biosynthesis. FAS is a large **multienzyme complex**. In mammals, FAS contains two subunits, each containing multiple enzyme activities. In bacteria and plants, individual proteins, which associate into a large complex, catalyze the individual steps of the synthesis scheme.

Initiation

Fatty acid synthesis starts with acetyl-CoA, and the chain grows from the "tail end" so that carbon 1 and the alpha-carbon of the complete fatty acid are added last. The first reaction is the transfer of the acetyl group to a pantothenate group of **acyl carrier protein** (ACP), a region of the large mammalian FAS protein. (The acyl carrier protein is a small, independent peptide in bacterial FAS, hence the name.) The pantothenate group of ACP is the same as is found on Coenzyme A, so the transfer requires no energy input:

Acetyl~S-CoA + HS-ACP → HS-CoA + Acetyl~S-ACP

In the preceding reaction, the S and SH refer to the thio group on the end of Coenzyme A or the pantothenate groups. The ~ is a reminder that the bond between the carbonyl carbon of the acetyl group and the thio group is a "high energy" bond (that is, the activated acetyl group is easily donated to an acceptor). The second reaction is another transfer, this time, from the pantothenate of the ACP to cysteine sulfhydral (–SH) group on FAS.

Acetyl~ACP + HS-FAS → HS-ACP + Acetyl~S-FAS

Elongation

The pantothenate –SH group is now ready to accept a malonyl group from malonyl-CoA:

Figure 2-1

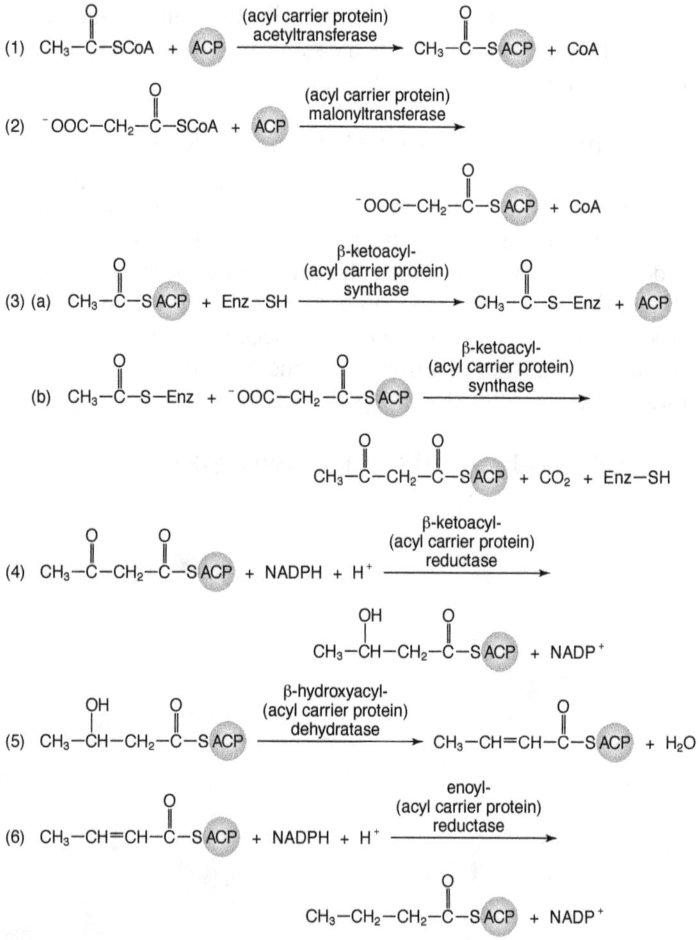

$$\text{(1)}\quad CH_3{-}\overset{O}{\overset{\|}{C}}{-}SCoA + ACP \xrightarrow[\text{acetyltransferase}]{\text{(acyl carrier protein)}} CH_3{-}\overset{O}{\overset{\|}{C}}{-}S\,ACP + CoA$$

$$\text{(2)}\quad {}^{-}OOC{-}CH_2{-}\overset{O}{\overset{\|}{C}}{-}SCoA + ACP \xrightarrow[\text{malonyltransferase}]{\text{(acyl carrier protein)}} {}^{-}OOC{-}CH_2{-}\overset{O}{\overset{\|}{C}}{-}S\,ACP + CoA$$

$$\text{(3) (a)}\quad CH_3{-}\overset{O}{\overset{\|}{C}}{-}S\,ACP + Enz{-}SH \xrightarrow[\text{synthase}]{\beta\text{-ketoacyl-(acyl carrier protein)}} CH_3{-}\overset{O}{\overset{\|}{C}}{-}S{-}Enz + ACP$$

$$\text{(b)}\quad CH_3{-}\overset{O}{\overset{\|}{C}}{-}S{-}Enz + {}^{-}OOC{-}CH_2{-}\overset{O}{\overset{\|}{C}}{-}S\,ACP \xrightarrow[\text{synthase}]{\beta\text{-ketoacyl-(acyl carrier protein)}} CH_3{-}\overset{O}{\overset{\|}{C}}{-}CH_2{-}\overset{O}{\overset{\|}{C}}{-}S\,ACP + CO_2 + Enz{-}SH$$

$$\text{(4)}\quad CH_3{-}\overset{O}{\overset{\|}{C}}{-}CH_2{-}\overset{O}{\overset{\|}{C}}{-}S\,ACP + NADPH + H^+ \xrightarrow[\text{reductase}]{\beta\text{-ketoacyl-(acyl carrier protein)}} CH_3{-}\overset{OH}{\overset{|}{C}H}{-}CH_2{-}\overset{O}{\overset{\|}{C}}{-}S\,ACP + NADP^+$$

$$\text{(5)}\quad CH_3{-}\overset{OH}{\overset{|}{C}H}{-}CH_2{-}\overset{O}{\overset{\|}{C}}{-}S\,ACP \xrightarrow[\text{dehydratase}]{\beta\text{-hydroxyacyl-(acyl carrier protein)}} CH_3{-}CH{=}CH{-}\overset{O}{\overset{\|}{C}}{-}S\,ACP + H_2O$$

$$\text{(6)}\quad CH_3{-}CH{=}CH{-}\overset{O}{\overset{\|}{C}}{-}S\,ACP + NADPH + H^+ \xrightarrow[\text{reductase}]{\text{enoyl-(acyl carrier protein)}} CH_3{-}CH_2{-}CH_2{-}\overset{O}{\overset{\|}{C}}{-}S\,ACP + NADP^+$$

Note that at this point, the FAS has two activated substrates, the acetyl group bound on the cysteine –SH and the malonyl group bound on the pantothenate –SH. Transfer of the 2-carbon acetyl unit from Acetyl~S-cysteine to malonyl-CoA has two features:

- Release of the CO_2 group of malonyic acid that was originally put on by acetyl-CoA carboxylase
- Generation of a 4-carbon **β-keto acid derivative**, bound to the pantothenate of the ACP protein

The ketoacid is now reduced to the methylene (CH_2) state in a three-step reaction sequence.

1. Reduction by NADPH to form the **β-hydroxy acid derivative**:

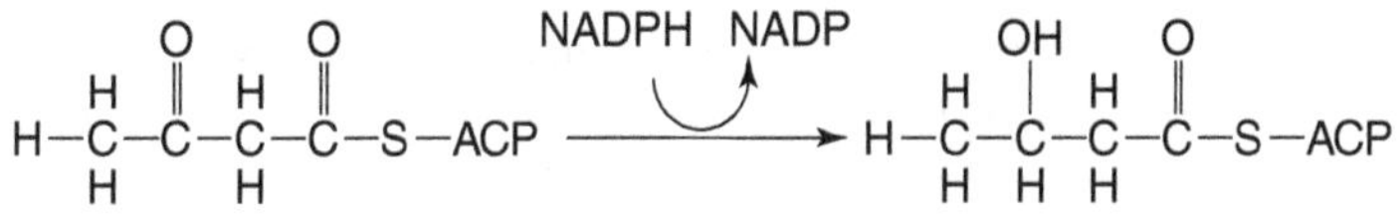

$$H{-}\underset{H}{\overset{H}{C}}{-}\overset{O}{\overset{\|}{C}}{-}\underset{H}{\overset{H}{C}}{-}\overset{O}{\overset{\|}{C}}{-}S{-}ACP \xrightarrow{NADPH \;\; NADP} H{-}\underset{H}{\overset{H}{C}}{-}\underset{H}{\overset{OH}{C}}{-}\underset{H}{\overset{H}{C}}{-}\overset{O}{\overset{\|}{C}}{-}S{-}ACP$$

2. Dehydration, that is, the removal of water to make a **trans-double bond**:

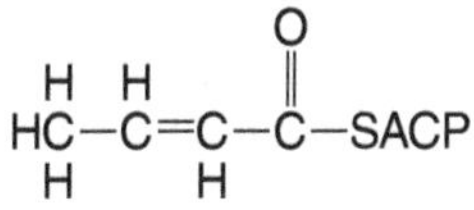

$$\underset{H}{\overset{H}{HC}}{-}\overset{H}{C}{=}\underset{H}{C}{-}\overset{O}{\overset{\|}{C}}{-}SACP$$

3. Reduction by NADPH to make the **saturated fatty acid**:

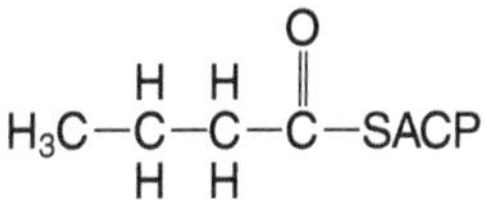

$$H_3C{-}\underset{H}{\overset{H}{C}}{-}\underset{H}{\overset{H}{C}}{-}\overset{O}{\overset{\|}{C}}{-}SACP$$

The elongated 4-carbon chain is now ready to accept a new 2-carbon unit from malonyl-CoA. The 2-carbon unit, which is added to the growing fatty acid chain, becomes carbons 1 and 2 of hexanoic acid (6-carbons).

Release

The cycle of transfer, elongation, reduction, dehydration, and reduction continues until **palmitoyl-ACP** is made. Then the **thioesterase** activity of the FAS complex releases the 16-carbon fatty acid palmitate from the FAS.

Note that fatty acid synthesis provides an extreme example of the phenomenon of **metabolic channeling**: neither free fatty acids with more than four carbons nor their CoA derivatives can directly participate in the synthesis of palmitate. Instead they must be broken down to acetyl-CoA and reincorporated into the fatty acid.

Fatty acids are generated cytoplasmically while acetyl-CoA is made in the mitochondrion by pyruvate dehydrogenase.This implies that a **shuttle system** must exist to get the acetyl-CoA or its equivalent out of the mitochondrion. The shuttle system operates in the following way: Acetyl-CoA is first converted to citrate by citrate synthase in the TCA-cycle reaction. Then citrate is transferred out of the mitochondrion by either of two carriers, driven by the electroosmotic gradient: either a citrate/phosphate antiport or a citrate/malate antiport as shown in Figure 2-2.

Figure 2-2

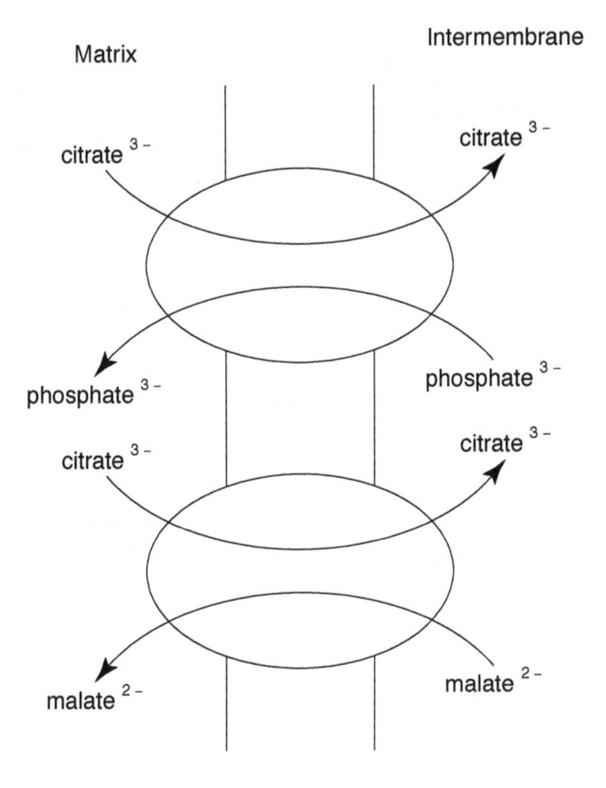

After it is in the cytosol, citrate is cleaved to its 2- and 4-carbon components by **citrate lyase** to make acetyl-CoA and oxaloacetate. Citrate lyase requires ATP.

Fatty acid biosynthesis (and most biosynthetic reactions) requires NADPH to supply the reducing equivalents. Oxaloacetate is used to generate NADPH for biosynthesis in a two-step sequence. The first step is the malate dehydrogenase reaction found in the TCA cycle. This reaction results in the formation of NAD from NADH (the NADH primarily comes from glycolysis). The malate formed is a substrate for the malic enzyme reaction, which makes pyruvate, CO_2, and NADPH. Pyruvate is transported into the mitochondria where pyruvate carboxylase uses ATP energy to regenerate oxaloacetate.

Palmitate is the starting point for other fatty acids that use a set of related reactions to generate the modified chains and head groups of the lipid classes. Microsomal enzymes primarily catalyze these chain modifications. Desaturation uses O_2 as the ultimate electron acceptor to introduce double bonds at the nine, six, and five positions of an acyl-CoA.

Elongation is similar to synthesis of palmitate because it uses malonyl-CoA as an intermediate. See Figure 2-3.

Figure 2-3

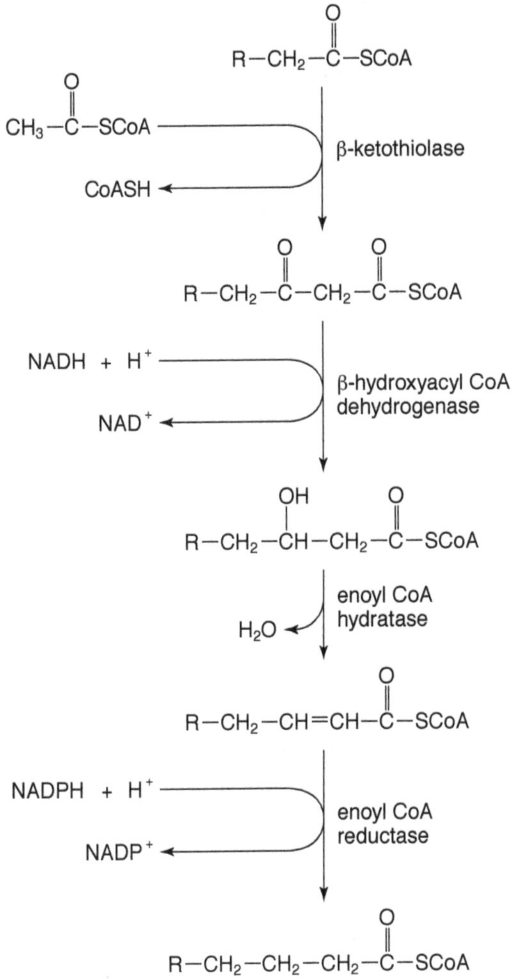

$$R{-}CH_2{-}\overset{O}{\overset{\|}{C}}{-}SCoA$$

$CH_3{-}\overset{O}{\overset{\|}{C}}{-}SCoA$ → CoASH

β-ketothiolase

$$R{-}CH_2{-}\overset{O}{\overset{\|}{C}}{-}CH_2{-}\overset{O}{\overset{\|}{C}}{-}SCoA$$

$NADH + H^+$ → NAD^+

β-hydroxyacyl CoA dehydrogenase

$$R{-}CH_2{-}\overset{OH}{\overset{|}{C}H}{-}CH_2{-}\overset{O}{\overset{\|}{C}}{-}SCoA$$

→ H_2O

enoyl CoA hydratase

$$R{-}CH_2{-}CH{=}CH{-}\overset{O}{\overset{\|}{C}}{-}SCoA$$

$NADPH + H^+$ → $NADP^+$

enoyl CoA reductase

$$R{-}CH_2{-}CH_2{-}CH_2{-}\overset{O}{\overset{\|}{C}}{-}SCoA$$

Synthesis of Triacylglycerols

Glycerol accepts fatty acids from acyl-CoAs to synthesize glycerol lipids. Glycerol phosphate comes from glycolysis—specifically from the reduction of dihydroxyacetone phosphate using NADH as a cofactor. Then the glycerol phosphate accepts two fatty acids from fatty acyl-CoA. The fatty acyl-CoA is formed by the expenditure of two high-energy phosphate bonds from ATP.

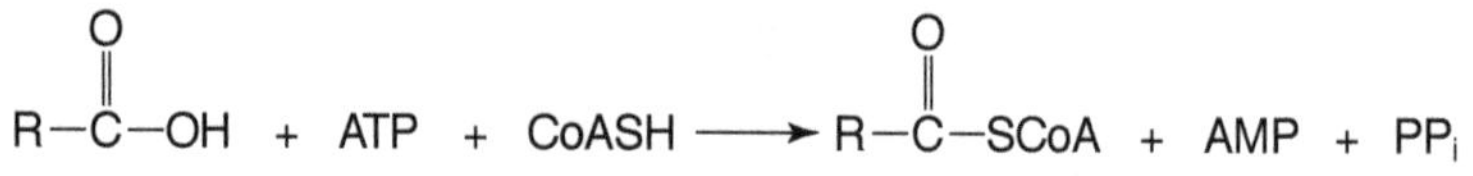

$$R{-}\overset{\overset{\displaystyle O}{\|}}{C}{-}OH + ATP + CoASH \longrightarrow R{-}\overset{\overset{\displaystyle O}{\|}}{C}{-}SCoA + AMP + PP_i$$

Fatty acyl-CoA is the donor of the fatty acyl group to the two *nonphosphorylated* positions of glycerol phosphate to make a *phosphatidic acid.*

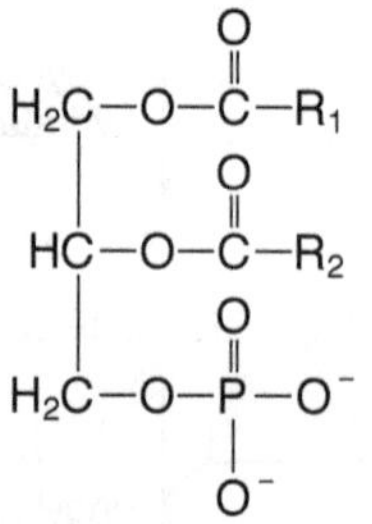

$$\begin{array}{l} H_2C{-}O{-}\overset{\overset{\displaystyle O}{\|}}{C}{-}R_1 \\ \;| \\ HC{-}O{-}\overset{\overset{\displaystyle O}{\|}}{C}{-}R_2 \\ \;| \\ H_2C{-}O{-}\overset{\overset{\displaystyle O}{\|}}{\underset{\underset{\displaystyle O^-}{|}}{P}}{-}O^- \end{array}$$

The third fatty acid can be added after the removal of the phosphate of the phosphatidic acid. This scheme results in a triacylglycerol, although other phosphatidic acids can be used as precursors to various membrane lipids.

Cholesterol Biosynthesis and its Control

Despite a lot of bad press, cholesterol remains an essential and important biomolecule in animals. As much as half of the membrane lipid in a cellular membrane is cholesterol, where it helps maintain constant fluidity and electrical properties. Cholesterol is especially prominent in membranes of the nervous system.

Cholesterol also serves as a precursor to other important molecules. Bile acids aid in lipid absorption during digestion. Steroid hormones all derive from cholesterol, including the adrenal hormones that maintain fluid balance; Vitamin D, which is an important regulator of calcium status; and the male and female sex hormones. Although humans wouldn't survive in one sense or another without cholesterol metabolites, cholesterol brings with it some well-known side effects. Doctors find cholesterol derivatives, being essentially insoluble in water, in the deposits (plaque) that characterize diseased arteries.

Isoprenoid Compounds

Cholesterol is synthesized from acetyl-CoA in the liver. Cholesterol and a number of natural products from plants (including rubber) are **isoprenoid** compounds. The isoprenoid unit is a 5-carbon structure.

```
        C
        |
C—C—C—C
```

Isoprenoid compounds are synthesized from a common intermediate, **mevalonic acid**. Mevalonate is synthesized from acetyl-CoA and then serves as the precursor to isoprenoid units.

acetyl-CoA → mevalonate

The key enzyme in this pathway is HMG-CoA reductase in connection with ketone body formation. The reactions leading to HMG-CoA are shared with that pathway.

Acetyl-CoA can be derived directly from metabolism, especially lipid degradation, or by the acetate thiokinase reaction:

$$H_3C-\overset{\overset{\displaystyle O}{\|}}{C}-OH + ATP + CoASH \longrightarrow H_3C-\overset{\overset{\displaystyle O}{\|}}{C}-SCoA + AMP + PP_i$$

Two acetyl-CoA molecules condense to form acetoacetyl-CoA.

$$H_3C-\overset{\overset{\displaystyle O}{\|}}{C}-SCoA + H_3C-\overset{\overset{\displaystyle O}{\|}}{C}-SCoA \longrightarrow H_3C-\overset{\overset{\displaystyle O}{\|}}{C}-CH_2-\overset{\overset{\displaystyle O}{\|}}{C}-SCoA + CoASH$$

Finally, the enzyme 3-hydroxy-3-methylglutaryl-CoA synthase (HMG-CoA synthase) adds a third acetyl-CoA.

$$\begin{array}{l} O{=}C-SCoA \\ \quad | \\ \quad CH_2 \\ \quad | \\ O{=}C \\ \quad | \\ \quad CH_3 \end{array} \quad H_3C-\overset{\overset{\displaystyle O}{\|}}{C}-SCoA \xrightarrow{H_2O} \begin{array}{l} O{=}C-SCoA \\ \quad\;\; | \\ \quad\;\; CH_2 \\ \quad\;\; | \\ HO-C-CH_3 \\ \quad\;\; | \\ \quad\;\; CH_2-\overset{\overset{\displaystyle O}{\|}}{C}-OH \end{array} + CoASH$$

HMG CoA Reductase

HMG-CoA reductase is the committed and therefore the regulatory step in cholesterol biosynthesis. If HMG-CoA is reduced to mevalonate, cholesterol is the only product that can result. The reduction is a two-step reaction, which releases the Coenzyme A cofactor and converts the thiol-bound carboxylic group of HMG-CoA to a free alcohol. Two NADPH molecules supply the reducing equivalents because the thioester must first be reduced to the level of an aldehyde and then to an alcohol.

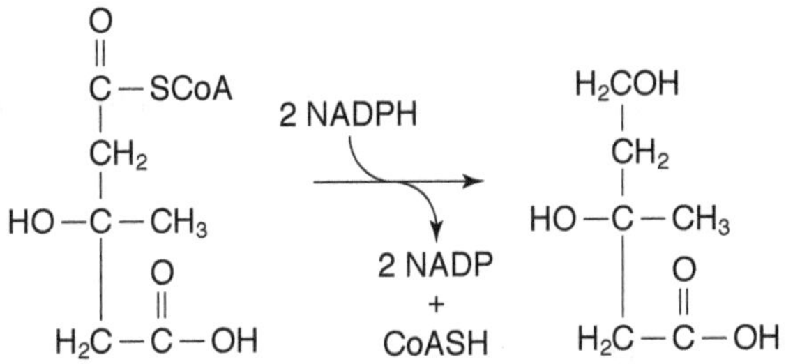

$$\begin{array}{c} O \\ \| \\ C-SCoA \\ | \\ CH_2 \\ | \\ HO-C-CH_3 \\ | \\ H_2C-C(=O)-OH \end{array} \xrightarrow[\text{2 NADP} + \text{CoASH}]{\text{2 NADPH}} \begin{array}{c} H_2COH \\ | \\ CH_2 \\ | \\ HO-C-CH_3 \\ | \\ H_2C-C(=O)-OH \end{array}$$

Mevalonate Squalene

Mevalonate molecules are condensed to a 30-carbon compound, squalene. The alcohol groups of mevalonate are first phosphorylated. Then they multiply phosphorylated mevalonate decarboxylates to make the two compounds isopentenyl pyrophosphate (IPP) and dimethylallyl pyrophosphate (DMAPP).

mevalonate → phosphomevalonate → pyrophosphomevalonate

This is a simple sequence of two phosphorylations using ATP as the donor:

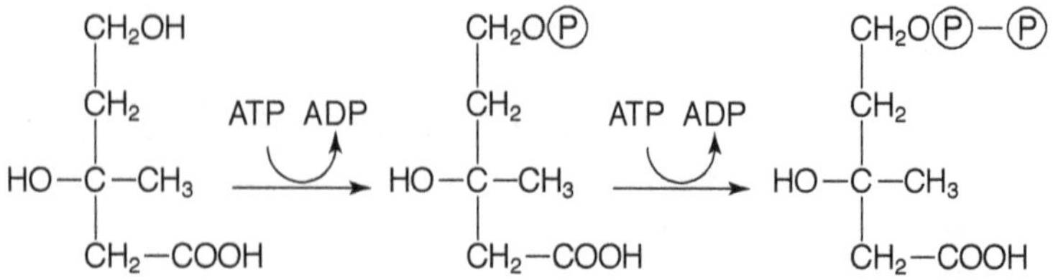

pyrophosphomevalonate ⇌ isopentenyl pyrophosphate ⇌ dimethylallyl pyrophosphate.

First, the other hydroxyl group of mevalonate accepts a phosphate from ATP. The resulting compound rearranges in an enzyme-catalyzed reaction, eliminating both CO_2 and phosphate. The 5-carbon compound that results, IPP, is rapidly isomerized with DMAPP.

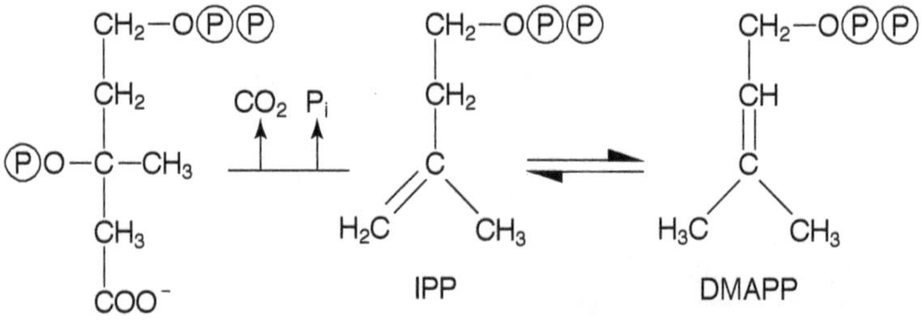

geranyl and farnesyl pyrophosphates → squalene

In plants and fungi, IPP and DMAPP are the precursors to many so-called **isoprenoid** compounds, including natural rubber. In animals, they are mainly precursors to sterols, such as cholesterol. The first step is condensation of one of each to geranyl pyrophosphate, which then condenses with another molecule of IPP to make **farnesyl pyrophosphate**. Some important membrane-bound proteins have a farnesyl group added on to them; however, the primary fate of farnesyl pyrophosphate is to accept a pair of electrons from NADPH and condense with another molecule of itself to release both pyrophosphate groups.

$$(H_3C)_2C{=}CH{-}CH_2O\text{Ⓟ}\text{Ⓟ} \; + \; CH_2{=}C(CH_3){-}CH_2{-}CH_2O\text{Ⓟ}\text{Ⓟ} \longrightarrow$$

$$(H_3C)_2C{=}CH{-}CH_2{-}CH_2{-}C(CH_3){=}CH{-}CH_2O\text{Ⓟ}\text{Ⓟ} \; + \; PP_i \xrightarrow{\text{DMAPP} \; \to \; \text{PP}}$$

$$(H_3C)_2C{=}CH{-}CH_2{-}CH_2{-}C(CH_3){=}CH{-}CH_2{-}CH_2{-}C(CH_3){=}CH{-}CH_2O\text{Ⓟ}\text{Ⓟ} \; (\times 2)$$

The resulting 30-carbon compound is **squalene**; it folds into a structure that closely resembles the structure of the steroid rings, although the rings are not closed yet.

$$(H_3C)_2C{=}CH{-}CH_2{-}CH_2{-}C(CH_3){=}CH{-}CH_2{-}CH_2{-}C(CH_3){=}CH{-}CH_2{-}CH_2{-}CH{=}C(CH_3){-}CH_2{-}CH_2{-}CH{=}C(CH_3){-}CH_2{-}CH_2{-}CH{=}C(CH_3){-}CH_3$$

Squalene → Lanosterol

The first recognizable steroid ring system is **lanosterol**; it is formed first by the epoxidation of the double bond of squalene that was originally derived from a DMAPP through farnesyl pyrophosphate, and then by the cyclization of squalene epoxide. The enzyme that forms the epoxide uses NADPH to reduce molecular oxygen to make the epoxide. See Figure 2-4.

Figure 2-4

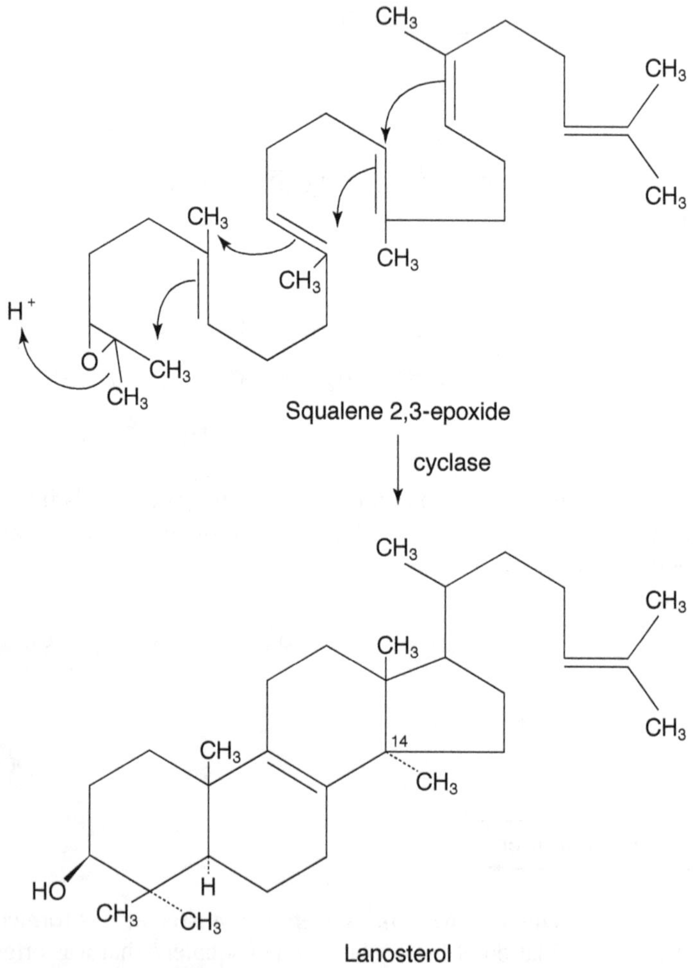

Squalene 2,3-epoxide

cyclase

Lanosterol

Lanosterol → Cholesterol

This sequence of reactions is incompletely understood but involves numerous oxidations of carbon groups, for example, the conversion of methyl groups to carboxylic acids, followed by decarboxylation. The end product, cholesterol, is the precursor to cholesterol esters in the liver and is transported to the peripheral tissues where it is a precursor to membranes (all cells), bile salts (liver), steroid hormones (adrenals and reproductive tissues), and vitamin D (skin, then liver, and finally kidney).

Cholesterol Transport, Uptake, and Control

Cholesterol is exported to the peripheral tissues in LDL and VLDL (see Chapter 1). About 70 percent of the cholesterol molecules in LDL are **esterified** with a fatty acid (for example, palmitate) on the OH group (at Carbon 3; see Figure 2-5). Cells take up cholesterol from the LDL by means of **LDL receptors** in the outer cell membrane.

Figure 2-5

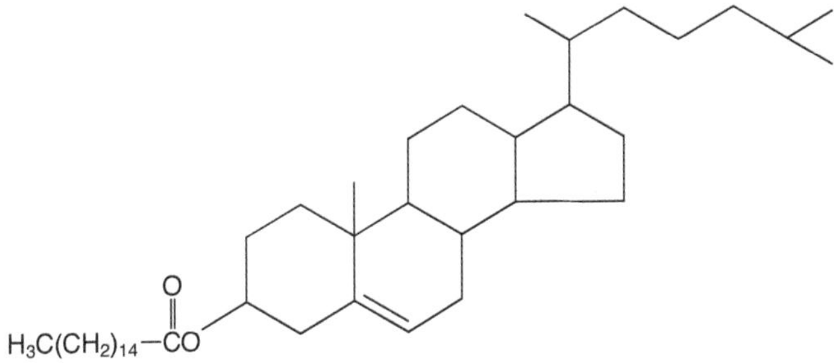

$H_3C(CH_2)_{14}-\overset{O}{\overset{\|}{C}}O$

cholesterol ester

The pathway for uptake involves several steps, including the following:

1. The assembly of the receptor-LDL complexes into a **coated pit** on the cell surface.
2. The pit folds into a spherical **endosome**, which is a small vesicle of cell membrane with receptor-LDL complexes on the inside.
3. The endosome fuses with a lysosome containing a large number of degradative enzymes and a low pH on the inside.
4. The receptors separate from the endosome-lysosome and return to the cell surface.
5. The cholesterol esters are hydrolyzed to free cholesterol.
6. The free cholesterol inhibits the synthesis and/or causes the degradation of HMG-CoA reductase and of LDL receptor. This last step ensures that more cholesterol will not be taken up or made than is needed.
7. Cholesterol is re-esterified with a fatty acid for storage inside the cell.

Figure 2-6

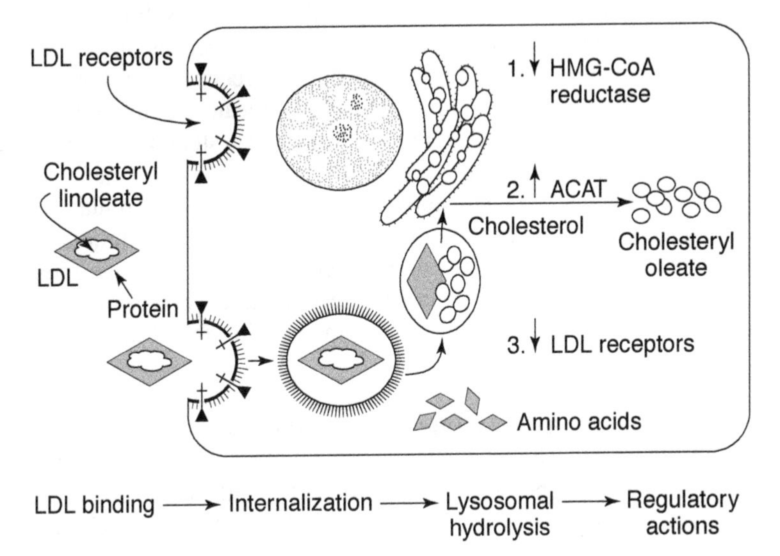

A close connection exists between the regulation of cholesterol biosynthesis and uptake. When HMG-CoA reductase is inhibited, the cell responds by synthesizing more LDL receptors to ensure the uptake of cholesterol from the serum. When cholesterol is present in a high enough concentration in the cell, LDL receptors are not exported to the cell surface, an example of the phenomenon of **down regulation**.

The tight regulation of cholesterol metabolism helps explain the pathology of coronary artery disease, a major killer in developed countries. Clearly, diet affects coronary artery disease: Individuals with high intake of saturated fat and cholesterol are most at risk. Furthermore, a high serum concentration of LDL cholesterol is associated with an increased risk of coronary artery disease.

The capability of LDL receptors to *remove* LDL cholesterol from the circulation can rationalize these clinical observations. If little cholesterol is available in the diet, the cells of the peripheral tissues respond by up-regulating the number of LDL receptors on the cell surface. The higher concentration of receptors means that more of the cholesterol will be removed from the circulatory system. Because the inappropriate deposition of cholesterol is a major contributor to blocked arteries, if the cholesterol is removed from the circulation, less risk of blockage exists. On the other hand, if a large amount of cholesterol exists in the diet, and the cells have enough for their needs, they will synthesize fewer LDL receptors, less cholesterol will be removed from the circulatory system, and the risk of artery disease increases further.

Several therapies are used to treat individuals with high serum cholesterol levels. The first is a low-fat, low-cholesterol diet. If the diet provides less cholesterol, then the cells will synthesize more LDL receptors to meet their needs, which means that more cholesterol will be removed from the circulation. The second therapy—which goes hand in hand with the first—is to decrease the reabsorption of bile acids in the gut, for example, by increasing the amount of soluble fiber in the diet or by administering synthetic bile acid binders. Bile acids bind to these agents. Fiber and the synthetic binders cannot be absorbed by the intestines and are excreted, carrying the bile acids with them. Excess cholesterol is then converted to bile acids and ultimately excreted. The third method is to inhibit HMG-CoA synthesis with any of several drugs on the market. HMG-CoA reductase carries out the committed step of cholesterol biosynthesis. Inhibiting this enzyme decreases the amount of cholesterol synthesized intracellularly, and the cells compensate by increasing the number of LDL receptors on the cell surface. This helps remove LDL cholesterol from the circulation.

CHAPTER 3
PHOTOSYNTHESIS

Metabolic Oxidation and Reduction

Metabolic energy derives from processes of oxidation and reduction. When energy is consumed in a process, chemical energy is made available for synthesis of ATP as one atom gives up electrons (becomes oxidized) and another atom accepts electrons (becomes reduced). For example, observe the following aerobic metabolism of glucose.

$$C_6H_{12}O_6 + 6\ O_2 \rightarrow 6\ CO_2 + 6\ H_2O + \text{energy}$$

The carbon in glucose moves from an oxidation state of zero to an oxidation state of +4. Concurrently, elemental oxygen moves from its oxidation state of zero to an oxidation state of –2 during the process.

Anaerobic catabolic reactions are similar, although the electron acceptor isn't oxygen. The next example shows the fermentation of glucose to lactic acid.

$$C_6H_{12}O_6 \rightarrow 2\ H_3CCHOHCO_2H + \text{energy}$$

In this case, one carbon (the methyl carbon of lactic acid) is reduced from the zero oxidation state to –3 while another carbon (the carboxyl carbon of lactic acid) gives up electrons and goes from an oxidation state of zero to +3. In this example, the electron acceptor and electron donor are located on the same molecule, but the principle remains the same: One component is oxidized and one is reduced at the same time.

Reactions that run in the opposite direction of the preceding ones, particularly the first, must exist. Glucose must be made from inorganic carbon—that is, CO_2. More generally, reducing equivalents and energy must be available to carry out the synthetic reaction.

$6\ CO_2 + 12\ RH_{red} + energy \rightarrow C_6H_{12}O_6 + 12\ R_{ox} + 3$ Oxygen (or equivalent)

The general reaction accounts for the fact that in some systems, something other than water supplies the reducing equivalents. For example, bacteria living in deep-sea thermal vents can apparently use hydrogen sulfide (H_2S) as a source of reducing equivalents to synthesize glucose from carbon dioxide dissolved in the seawater.

Overall Process of Photosynthesis

The best-understood reaction for the synthesis of glucose, and probably the most important quantitatively, is photosynthesis. Photosynthesis converts carbon from carbon dioxide to glucose with reducing equivalents supplied from water and energy supplied from light.

$$6\ CO_2 + 6\ H_2O + \text{light energy} \rightarrow C_6H_{12}O_6 + 6\ O_2$$

The energy in light is dependent on its wavelength, and is given by the following relationship.

$$E = h\nu = \frac{hc}{\lambda}$$

The Greek letter nu, ν, stands for the frequency of the light, h is a constant called Planck's constant, c is the speed of light, and λ is the wavelength. In other words, the energy of light is inversely proportional to its wavelength. The longer the wavelength, the less energy it contains. In the visible spectrum, the highest-energy light is toward the blue or violet end, while the lowest-energy is to the red.

Two photosynthetic reactions

Photosynthesis involves two sets of chemical events, termed the **light** and **dark reactions**. This terminology is somewhat misleading, because the entire process of photosynthesis is regulated to take place when an organism absorbs visible light. The light reactions refer to the set of reactions in which the energy of absorbed light is used to generate **ATP** and reducing power (**NADPH**). The dark reactions use this reducing power and energy to fix carbon, that is, to convert carbon dioxide to glucose. Biochemically, converting CO_2 to glucose without light is possible if a supply of reducing equivalents and ATP are available.

The chloroplast

In higher plants, both the light and the dark reactions take place in the chloroplast, with each reaction set occurring in a different substructure. In electron micrographs, the chloroplast is seen as a series of membranes that come together to form **grana**, or grains, set in the **stroma**, or spread-out region as seen in Figure 3-1. Within the grana, the membranes stack upon each other in a disk-like arrangement called the **thylakoid**. Each region of the chloroplast is specialized to carry out a specific set of reactions. The light reactions occur in the grana and the dark reactions occur in the stroma.

Chlorophyll and the action spectrum of photosynthesis

The green color of the chloroplast (and therefore plants) comes from the chlorophyll that is stored in them. Chlorophyll is a **tetrapyrrole** ring system with a Mg2+ ion in the center, coordinated to the nitrogen of each pyrrole ring. The tetrapyrrole ring system is found as a bound cofactor (a prosthetic group) in many electron-carrying proteins, enzymes, and oxygen transporters. For example, tetrapyrroles are essential to the functioning of cytochrome c, various mixed function oxidases, and hemoglobin. Chlorophylls differ from other tetrapyrroles by possessing a long, branched **phytol** joined to the tetrapyrrole in an ether linkage. The phytol is an "anchor" to keep the chlorophyll inside the thylakoid membrane.

Figure 3-1

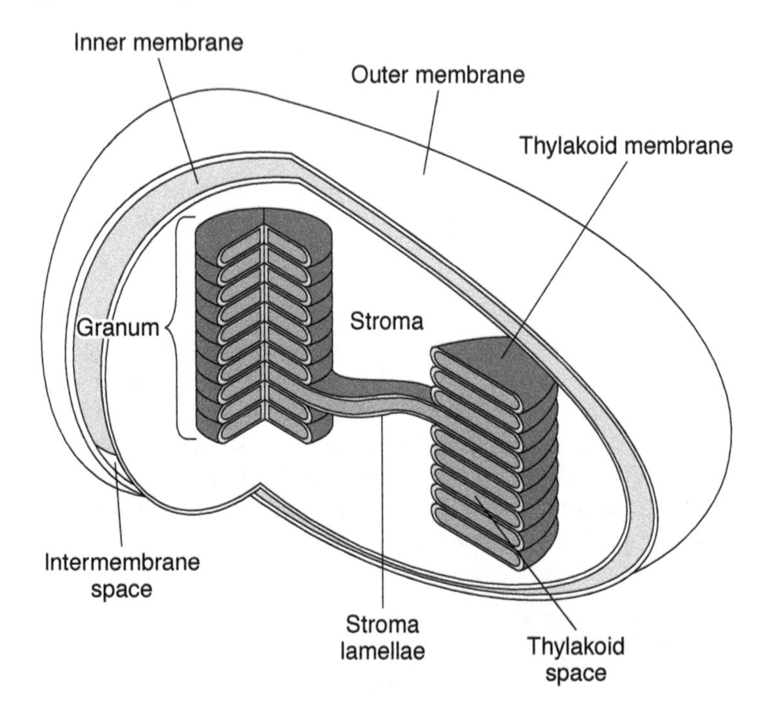

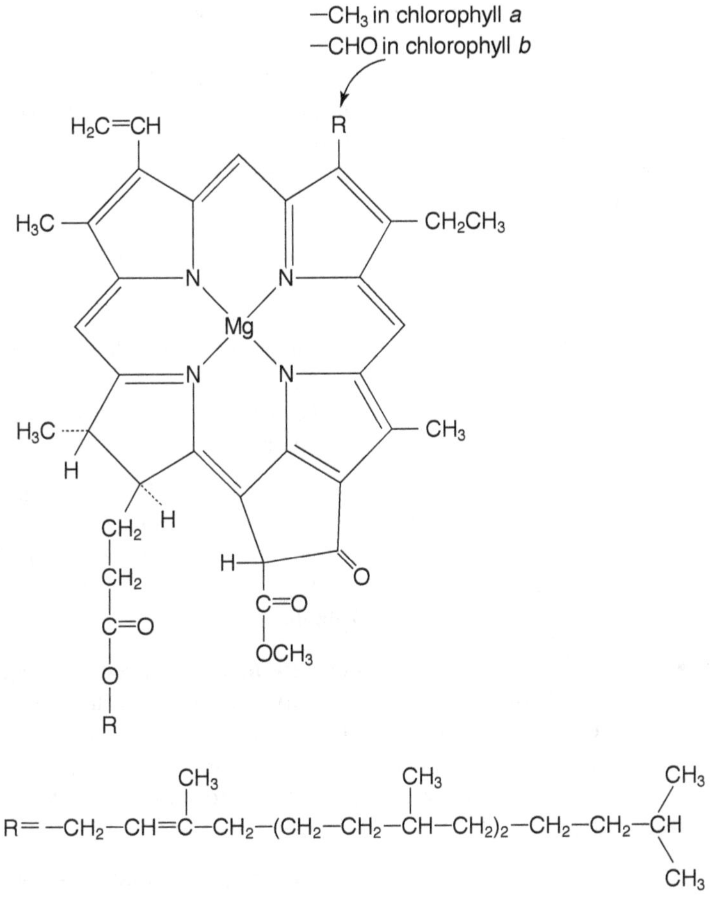

—CH_3 in chlorophyll *a*
—CHO in chlorophyll *b*

R

$H_2C{=}CH$

H_3C — — CH_2CH_3

N N

Mg

N N

H_3C — — CH_3

H

H

CH_2

CH_2

$C{=}O$

O

R

H— $C{=}O$ O

OCH_3

$$R = -CH_2-CH{=}C(CH_3)-CH_2-(CH_2-CH_2-CH(CH_3)-CH_2)_2-CH_2-CH_2-CH(CH_3)_2$$

Photosynthesis begins with the absorption of light in the thylakoid membrane. The energy of the light makes a difference in its effect on photosynthesis. The following considerations can help you understand this concept.

- The energy of a single photon of light is inversely proportional to its wavelength, with the visible region of the spectrum having less energy per photon than the ultraviolet region, and more than the infrared region. The energy of the visible spectrum increases from the red wavelengths through the blue and violet, according to the mnemonic ROY G. BIV (red, orange, yellow, green, blue, indigo, violet).
- Ultraviolet light, which has more energy than blue light, does not support photosynthesis. If it did reach the earth's surface, ultraviolet light would be energetic enough to break carbon-carbon bonds. The bond-breaking process would lead to a net loss of fixed carbon as biomolecules were broken apart. Fortunately, the ozone layer in the atmosphere absorbs enough UV radiation to prevent this from occurring.
- Chlorophyll comes in two varieties, chlorophyll a and chlorophyll b. Although the wavelengths at which they absorb light differ slightly, both absorb red and blue light. The chlorophyll reflects the other colors of light; the human eye sees these colors as green, the color of plants.
- Other pigments, called **antenna pigments**, or accessory pigments, absorb light at other wavelengths. The accessory pigments are responsible for the brilliant colors of plants in the autumn (in the Northern Hemisphere). The breakdown of chlorophyll allows us to see the colors of the accessory pigments.
- The antenna pigments and most chlorophyll molecules don't participate in the direct light reactions of photosynthesis. Instead they are part of the **light-harvesting complex**, which "funnels" the photons they capture to a **reaction center**, where the actual reactions of photosynthesis occur. All together, the light-harvesting complex is over 90 percent efficient—almost all the photons that fall on the chloroplast are absorbed and can provide energy for synthesis.
- Chlorophyll a and chlorophyll b participate in aspects of the light reaction; each must absorb a photon for the reaction to occur.

Light Reactions

The photosynthetic scheme begins when the pigments of the light-harvesting complex absorb a photon of visible light. The absorption of light **excites** the pigment to a higher energy state. The pigments transfer their excitation energy to the reaction center chlorophyll molecule. Two types of reaction centers exist, generally called **Photosystem I (PSI)** and **Photosystem II (PSII)**. Thus, two types of overall reactions make up photosynthesis—the light and dark reactions—while two more types of reactions compose the light reaction—those carried out by PSI and PSII. Absorption of a photon causes either type of chlorophyll to become more easily oxidized—that is, to give up an electron. The electron given up by either chlorophyll molecule flows through an electron transport chain, just as the electron given up by a mitochondrial cytochrome flows through the mitochondrial electron transport chain. Just as in mitochondrial electron transport, the flow of electrons leads to a proton gradient, which is used to synthesize ATP. Unlike mitochondrial electron transport, the terminal electron acceptor isn't oxygen but rather NADP.

Photosystem II

The photosystem pigment of PSII is a form of chlorophyll termed P680, because it is a **pigment** that absorbs light with a wavelength of **680 nanometers**. Absorption of a photon by P680 leads to the excited form of the pigment, called P680*. P680* but not ground-state P680 gives up an electron to another molecule, **plastiquinone**.

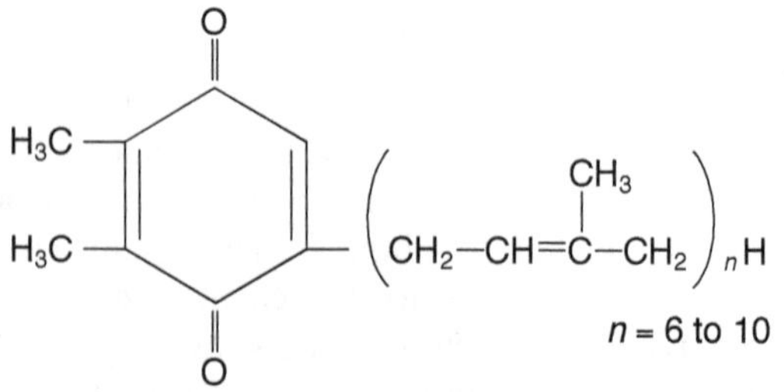

n = 6 to 10

Plastoquinone
(oxidized form, Q)

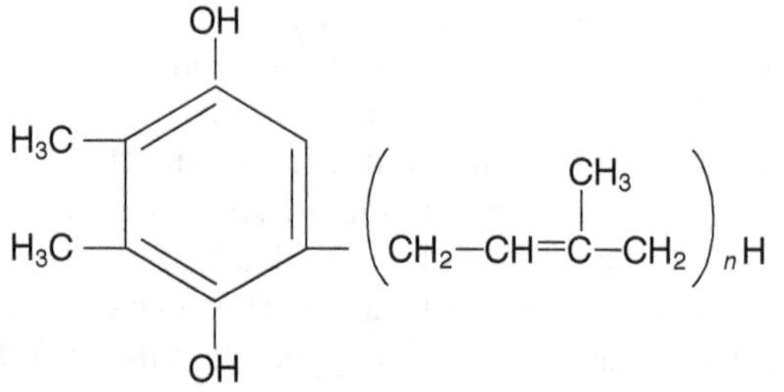

Plastoquinol
(reduced form, QH_2)

Another way of stating this relationship is that P680* has a more negative **reduction potential** than does P680. The following equation gives the free energy of the reduction reaction.

$$\Delta G^{\circ\prime} = -n \Im E^{\circ\prime}$$

The conversion of the P680 to P680* means that the free energy of its accepting an electron becomes more positive (the reaction is unfavored). A molecule that has a more positive free energy of reduction has a more negative free energy of oxidation. In other words, the P680* molecule is more easily oxidized than is P680.

The electron from P680* is transferred to a series of plastiquinone (PQ) derivatives, leaving behind an oxidized P680 molecule as shown in Figure 3-2. The reduction of plastoquinone is similar to that of Coenzyme Q in mitochondrial oxidation/reduction, in that PQ can accept either one or two electrons at a time. Plastiquinone molecules accept a proton (H+) from the stroma for each electron they accept. This leaves the stroma more basic than it was before, creating part of the gradient that will be used for ATP synthesis.

Figure 3-2

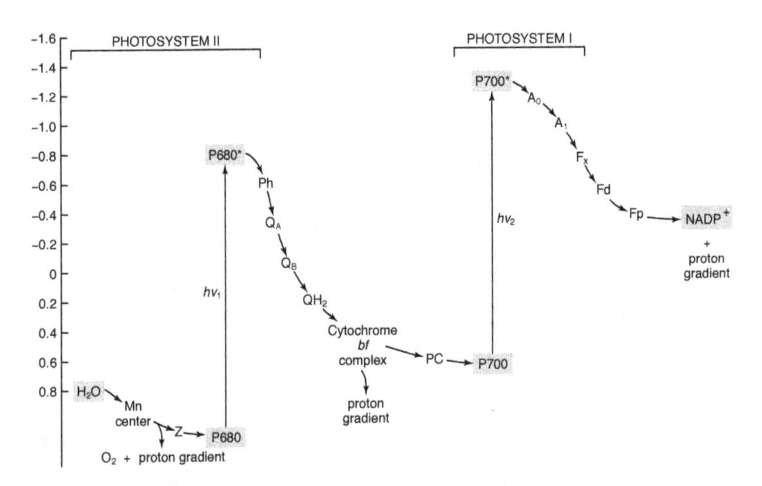

From the quinones, the electron is transferred to plastocyanin and then to cytochrome bf. The two H^+ ions (protons) left behind remain in the thylakoid lumen. As the electrons move down this electron transport chain, protons are pumped into the thylakoid lumen. Eventually the transported electron is given up to the oxidized **P700** chlorophyll of Photosystem I.

This movement of an electron to PSI from P680* leaves P680 in a non-excited, **oxidized state**. Oxidized P680 must be reduced to give up another electron. Hydrogen atoms derived from H_2O reduce it according to the reaction:

$$2\ H_2O + [\text{Manganese Center}]_{oxidized} \rightarrow$$

$$O_2 + [\text{Manganese Center}]_{reduced} + 4\ H^+$$

This oxidation transfers four electrons to the Manganese Center, a complex metalloprotein, which then donates the electrons through an intermediate to oxidized P680. The protons derived from water are transported into the thylakoid lumen. The protons pumped into the thylakoid lumen by PSII are used to make ATP through the action of **coupling factor**, in a mechanism similar to that of mitochondrial ATP synthesis.

Photosystem I

Just as the absorption of a photon of light converts P680 to P680*, which is a better reductant, so too does the absorption of light convert the chlorophyll of PSI to a species that gives up an electron more easily. P700* donates an electron to a series of mobile quinones and then to a ferredoxin protein. The ferredoxin reduces NADP to NADPH. This provides reducing power for conversion of CO_2 to carbohydrate.

$$2\ Fd_{reduced} + NADP^+ + H^+ \rightarrow 2\ Fd_{oxidized} + NADPH$$

After P700* donates the electron to the quinones, oxidized P700 accepts an electron from the cytochrome bf, which is the acceptor of the electron from Photosystem II. The reduced P700 is now poised to absorb a photon's energy and the cycle can start over.

Cyclic Electron Flow

Suppose the level of NADP is low and the level of NADPH is high. Because reduced ferredoxin donates its electron to NADP, the electrons could possibly "back up" in the two photosystems. In other words, all the electron carriers would be fully reduced and no photosynthesis could take place. This situation doesn't happen because an alternative pathway exists by which electrons can be donated to cytochrome bf instead of to NADP. The movement of electrons through this pathway pumps protons into the thylakoid lumen. These protons, along with the ones generated by the action of PSII, are used to make ATP.

Z-Scheme of Photosynthesis

The "Z-scheme" describes the oxidation/reduction changes during the light reactions of photosynthesis. The vertical axis in the figure represents the reduction potential of a particular species—the higher the position of a molecular species, the more negative its reduction potential, and the *more easily it donates electrons*. See Figure 3-3.

In the Z-scheme, electrons are removed from water (to the left) and then donated to the lower (non-excited) oxidized form of P680. Absorption of a photon excites P680 to P680*, which "jumps" to a more actively reducing species. P680* donates its electron to the quinone-cytochrome bf chain, with proton pumping. The electron from cytochrome bf is donated to PSI, converting P700 to P700*. This electron, along with others, is transferred to NADP, forming NADPH. Alternatively, this electron can go back to cytochrome bf in cyclic electron flow.

ATP Synthesis

The protein complex that carries out this reaction is called **coupling factor** or, more accurately, **ATP synthase**. ATP synthase is a complex of several proteins, shaped like a mushroom, with the cap on the stromal side of the thylakoid disc and the stalk going through the thylakoid membrane. The actions of PSII and PSI pump protons into the lumen of the thylakoid. When chloroplasts are actively illuminated, the pH gradient produced by the two photosystems can be as much as four pH units, or a 10,000-fold difference in hydrogen ion concentration. These are the relative acidities of vinegar and tap water.

The membrane potential in chloroplasts is almost entirely composed of the ΔpH component. The thylakoid membrane is permeable to Mg^{2+} and Cl^- ions, so electrical neutrality is maintained. This differs from mitochondrial ATP synthesis where both a pH and an electrochemical potential exist. Because the chloroplast gradient is primarily ΔpH in nature, three protons must move across the membrane to synthesize an ATP, rather than the two that move during mitochondrial synthesis of a single ATP. The ATP and NADPH from synthesis are both formed on the **stromal** side of the thylakoid membrane. They are available for the fixation of CO_2, which occurs in the stroma. See Figure 3-3.

Figure 3-3

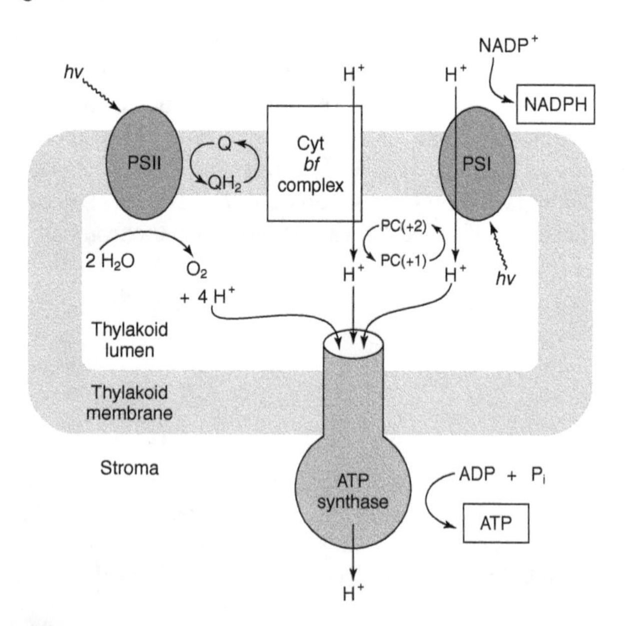

Carbohydrate Synthesis

The dark reactions use ATP and NADPH to convert CO_2 into carbohydrate. The first step is fixing CO_2 into organic carbon. The basic reaction is addition of CO_2 to a phosphorylated **acceptor**. This step requires no direct input of energy. Two types of plants exist, which use different acceptor molecules. In so-called **C-3 plants**, the acceptor is a 5-carbon, doubly phosphorylated acceptor, and two 3-carbon phosphorylated compounds are formed. In **C-4 plants**, the acceptor is phosphoenolpyruvate, and the carboxylation makes the 4-carbon acid oxaloacetatic acid and releases inorganic phosphate. The ATP and NADPH from the light reactions are used for making the acceptors and converting the first products into glucose.

C-3 Photosynthesis

This pathway is sometimes called the Calvin-Benson cycle, after the biochemists who elucidated it. The 5-carbon, doubly phosphorylated carbohydrate, **ribulose bisphosphate** is the acceptor for CO_2; the enzyme is called ribulose-bisphosphate carboxylase/oxygenase (called **Rubisco**).

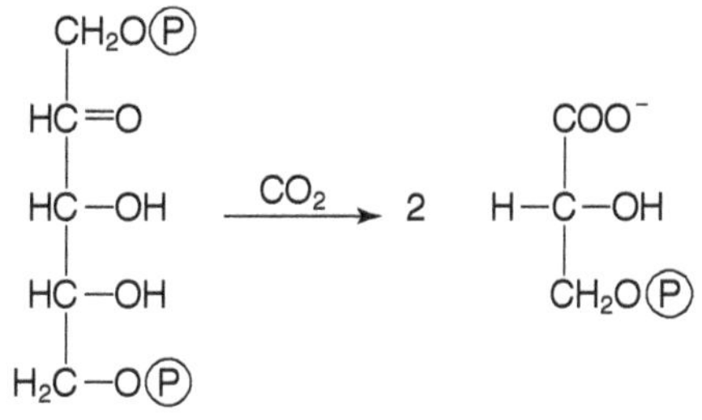

$$\begin{array}{l} CH_2O\text{Ⓟ} \\ | \\ HC{=}O \\ | \\ HC{-}OH \\ | \\ HC{-}OH \\ | \\ H_2C{-}O\text{Ⓟ} \end{array} \xrightarrow{CO_2} 2 \begin{array}{c} COO^- \\ | \\ H{-}C{-}OH \\ | \\ CH_2O\text{Ⓟ} \end{array}$$

The initial product of the reaction is unstable and quickly falls apart to yield two molecules of **3-phosphoglycerate**.

The rest of the Calvin cycle is involved in interconversion of carbohydrates to make glucose (or starch) and the regeneration of the ribulose-bisphosphate acceptor. The reactions are also found in the pathways for gluconeogenesis and the pentose phosphate shunt (see Volume 1, Chapters 10 and 12). The first step is the phosphorylation of 3-phosphoglycerate by the same reactions involved in gluconeogenesis.

3-phosphoglycerate + ATP → 1,3-bisphosphoglycerate + ADP

1,3-bisphosphoglycerate + NADPH + H^+ →
glyceraldehyde-3-phosphosphate + NADP + P_i

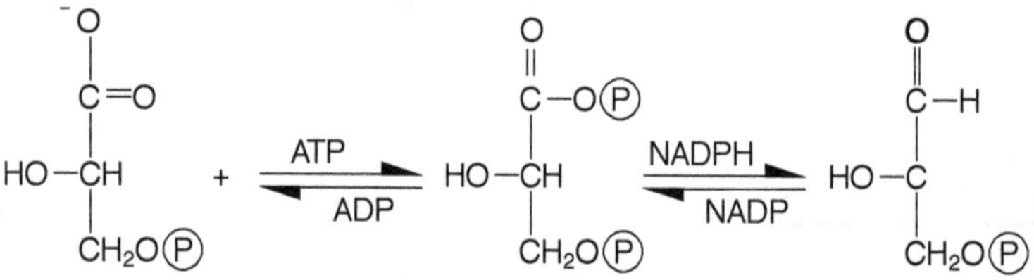

$$^{-}O\text{—}C(=O)\text{—}CH(OH)\text{—}CH_2O\text{Ⓟ} \; + \; \underset{ADP}{\overset{ATP}{\rightleftharpoons}} \; O=C(\text{—}O\text{Ⓟ})\text{—}CH(OH)\text{—}CH_2O\text{Ⓟ} \; \underset{NADP}{\overset{NADPH}{\rightleftharpoons}} \; O=C(\text{—}H)\text{—}C(OH)\text{—}CH_2O\text{Ⓟ}$$

The glyceraldehyde-3-phosphate can be converted into the 6-carbon sugar phosphate called fructose-6-phosphate by the reactions of triose phosphate isomerase, aldolase, and fructose bisphosphase.

glyceraldehyde-3-phosphate ⇄ dihydroxyacetone phosphate

glyceraldehyde-3-phosphate + dihydroxyacetone phosphate ⇄ fructose-1,6-bisphosphate

fructose-1,6-bisphosphate → fructose-6-phosphate + phosphate

Note that the last step is irreversible, just as it is in gluconeogenesis.

In shorthand, the reactions to this point are:

5 + 1 = 3 + 3 = 6

Regeneration of ribulose bisphosphate occurs by the same reactions that occur in the hexose-monophosphate shunt. The hexose monophosphate shunt interconverts 3-, 4-, 5-, 6- and 7-carbon sugar phosphates. In the formation of one molecule of glucose from CO_2, carrying out the previous reactions six times is necessary. Making the six molecules of ribulose-bisphosphate required for fixing six CO_2's occurs by the following sets of reactions:

- *Reaction 1: Transketolase transfers a 2-carbon unit*

fructose-6-phosphate + glyceraldehyde-3-phosphate $\rightleftharpoons$ erythrose-4-phosphate + xylulose-5-phosphate

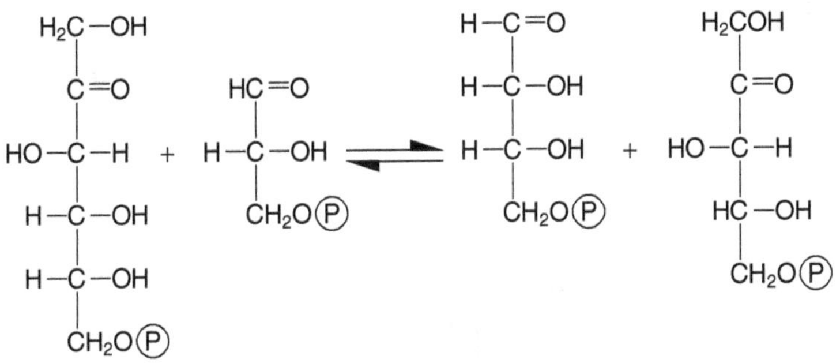

```
 H2C—OH                                  H—C=O            H2COH
    |                                       |                |
   C=O           HC=O                    H—C—OH             C=O
    |              |                        |                |
HO—C—H    +    H—C—OH     <=====>        H—C—OH    +     HO—C—H
    |              |                        |                |
 H—C—OH          CH2O(P)                  CH2O(P)           HC—OH
    |                                                        |
 H—C—OH                                                    CH2O(P)
    |
  CH2O(P)
```

- *Reaction 2: Transaldolase transfers a 3-carbon unit*

erythrose-4-phosphate + glyceraldehyde-3-phosphate $\rightleftharpoons$ sedoheptulose-1,7-bisphosphate

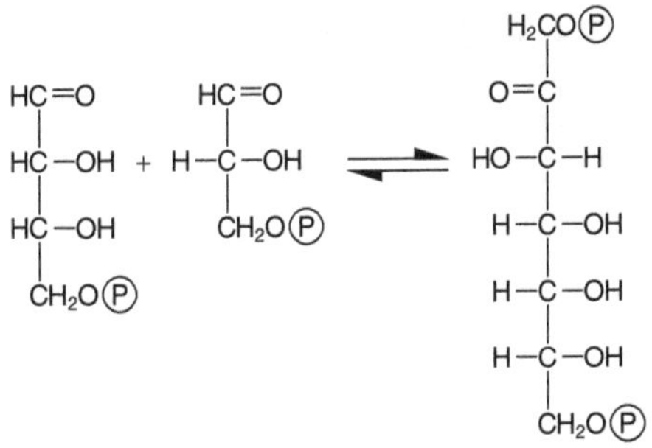

```
                                           H2CO(P)
                                             |
HC=O         HC=O                           O=C
 |             |                             |
HC—OH   +   H—C—OH        <=====>         HO—C—H
 |             |                             |
HC—OH        CH2O(P)                       H—C—OH
 |                                           |
CH2O(P)                                    H—C—OH
                                             |
                                           H—C—OH
                                             |
                                           CH2O(P)
```

- *Reaction 3: Transketolase transfers a 2-carbon unit*

sedoheptulose-1,7-bisphosphate → sedoheptulose-7-phosphate

sedoheptulose-7-phosphate + glyceraldehyde-3-phosphate ⇄ ribose-5-phosphate + xylulose-5-phosphate

```
    H2COH                                  H2C—OH          CH2OH
      |                                      |               |
      C=O       HC=O                       H—C=O           O=C
      |          |                           |               |
  HO—C—H    +   HC—OH      ⇄              H—C—OH    +   HO—C—H
      |          |                           |               |
   H—C—OH       CH2O(P)                    H—C—OH           HC—OH
      |                                      |               |
   H—C—OH                                  CH2O(P)         CH2O(P)
      |
   H—C—OH
      |
    CH2O(P)
```

Remembering all these reactions can be difficult. Thankfully, a useful mnemonic exists to help you keep track of the number of carbons in each of the three reactions:

Reaction 1: **6 + 3 ⇄ 4 + 5**

Reaction 2: **4 + 3 ⇄ 7**

Reaction 3: **7 + 3 ⇄ 5 + 5**

Summary: **6 + 3 + 3 + 3 → 5 + 5 + 5**

The 5-carbon sugar phosphates are interconverted by the action of **epimerase** and **isomerase** to yield ribulose-5-phosphate, which is phosphorylated by the enzyme **ribulose phosphate kinase** to make RuBP, the acceptor of CO_2.

Ribulose phosphate kinase is active only when a cystine disulfide on the enzyme is reduced to two cysteines. An electron carrier, thioredoxin, carries out this reduction, and is then itself reduced by electrons from NADPH. *Because the action of Photosystems I and II forms NADPH, this reduction ensures that ribulose bisphosphate is made only when enough light exists to support Photosynthesis*. In other words, the light and dark reactions are **coupled**.

Energetics of Photosynthesis

This coupling is necessary because photosynthesis consumes a large amount of energy. To carry out the synthesis of a single molecule of glucose, the previous set of reactions must be carried out six times at a minimum.

$$6\ CO_2 \rightarrow \text{glucose}$$

This requires the generation of six separate molecules of ribulose bisphosphate from ribulose phosphate, at the cost of one ATP each. Furthermore, two molecules of 1,3-bisphosphoglycerate must be made from the two 3-phosphoglycerates that are the initial product of each CO_2 fixation reaction. Conversion of each 1,3-bisphosphoglycerate requires an NADPH as well; therefore, two NADPH equivalents are consumed for each CO_2 fixed. Another way of saying this is that carbon is reduced from an oxidation number of –4 in CO_2 to an oxidation number of zero in carbohydrate (CH_2O). *Therefore, synthesis of one mole of glucose requires the input of 18 ATPs and 12 NADPHs*.

Generating one NADPH from NADP requires the transfer of two electrons through PSI and PSII. Transfer of one electron requires that a photon be absorbed by both PSI and PSII, so that the generation of 12 ATPs requires the absorption of $12 \times 2 \times 2 = 48$ photons. These 48 photons allow the synthesis of (just) 18 ATPs by coupling factor. Thus, the equation for photosynthesis can be written in two steps.

Light reactions:

48 photons + 18 ADP + 18 P_i + 12 NADP + 12 H_2O →
18 ATP + 12 NADPH + 12 H^+ + 6 O_2

Dark reactions:

6 CO_2 + 18 ATP + 12 NADPH + 12 H^+ →
$C_6H_{12}O_6$ + 18 ADP + 18 P_i + 12 NADP

The energy in 18 moles of photons of blue light is about 8000 kJ (~1800 kcal) and the energy required to synthesize a mole of glucose from CO_2 and water is 2870 kJ (684 kcal). This means that the maximal efficiency of photosynthesis is about 35 percent. Roughly one third of the available energy from sunlight is converted into glucose under optimal conditions. This assumes that no cyclic electron flow occurs, which would further lower the efficiency of photosynthesis. Although 35 percent may seem inefficient, this value is actually quite good—about the same efficiency as a diesel engine.

Photorespiration

While this efficiency is impressive, it also is rarely achieved. The difficulty is in the protein that carries out the first step of photosynthesis. Molecular oxygen, O_2, competes with CO_2 for the active site of ribulose bisphosphate carboxylase, leading to an oxidation and loss of the ribulose bisphosphate acceptor. This competition is apparently intrinsic to the enzyme, because attempts to increase the discrimination for CO_2 by genetic engineering have resulted in a less-active enzyme, which fixes CO_2 very poorly.

Oxidation of ribulose-1,5-bisphosphate by Rubisco produces a 3-carbon compound, 3-phosphoglycerate, and a 2-carbon compound, phosphoglycolate. Because carbon is oxidized, the process is termed **photorespiration**.

H_2CO(P)
C=O
H—C—OH + O_2 ⟶
H—C—OH
CH_2O(P)

H_2C—O—(P)
$^-$O—C=O phosphoglycolate
+
$^-$O—C=O
HCOH
CH_2O(P)

Photorespiration reduces the efficiency of photosynthesis for a couple of reasons. First, oxygen is added to carbon. In other words, the carbon is **oxidized**, which is the reverse of photosynthesis—the reduction of carbon to carbohydrate. Secondly, it is now necessary to resynthesize the ribulose bisphosphate and to reduce the phosphoglycolate.

The 3-phosphoglycerate from photorespiration can reenter the Calvin-Benson pathway, but the phosphoglycolate must be recycled to make a useful compound. This recycling takes place in a specialized organelle termed the **peroxisome**. Peroxisomes lie between chloroplasts and mitochondria in the plant cell and serve to pass the 2-carbon products of oxygenation on for further metabolism. In the chloroplast, the phosphoglycolate is dephosphorylated. Glycolate is transported to the peroxisome where molecular oxygen further oxidizes it to glyoxylate. The product is hydrogen peroxide, H_2O_2, (the term peroxisome comes from this product) which is rapidly broken down by catalase to water and oxygen.

The glyoxylate is amidated to the amino acid glycine in the peroxisome.

Glycine is then transported to the mitochondrial matrix where the conversion of two glycines to one serine occurs with the loss of CO_2 and NH_3 from the pool of fixed molecules. The serine is transported into the peroxisome, where it is deaminated to glycerate. The glycerate is transported back to the chloroplast, where it is phosphorylated to 3-phosphoglycerate for the Calvin-Benson cycle.

$$\mathrm{O{=}C{-}O^-\,|\,HC{=}O} \;+\; \mathrm{{}^-O{-}C{=}O\,|\,H_3{}^+N{-}CH\,|\,CH_2\,|\,CH_2\,|\,H_2N{-}C{=}O} \longrightarrow \mathrm{{}^-O{-}C{=}O\,|\,H_2C{-}NH_3{}^+} \;+\; \mathrm{{}^-O{-}C{=}O\,|\,H_3{}^+N{-}CH\,|\,CH_2\,|\,CH_2\,|\,{}^-O{-}C{=}O}$$

This set of reactions is very detrimental to the efficiency of photosynthesis. Oxygen is added to carbon, CO_2 is lost, energy is consumed, and ribulose bisphosphate is destroyed. For a plant to be able to increase the discrimination of Rubisco for CO_2 would obviously be advantageous, but that hasn't happened, either naturally or through the efforts of scientists. An increased concentration of CO_2 in the atmosphere may lead to increased photosynthesis and decreased photorespiration, but high CO_2 concentrations would also contribute to global warming (and the increased photosynthetic carbon fixation would not likely reduce the amount of CO_2 in any event).

Figure 4-4

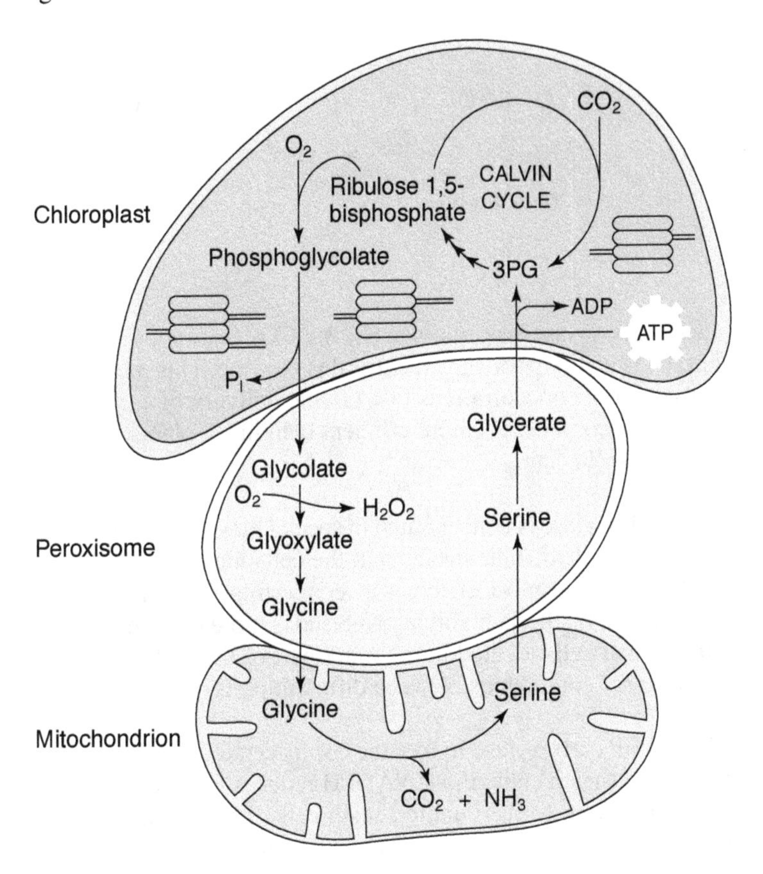

C4 plants

Photorespiration could be overcome if Rubisco could receive CO_2 more efficiently than by diffusion through the leaves. Various C4 plants, including grasses such as maize (corn) and bamboo, have evolved a bypass system for the delivery of CO_2 to Rubisco.

In these C4 plants, the enzyme phosphoenolpyruvate carboxylase first converts CO_2 to oxaloacetate.

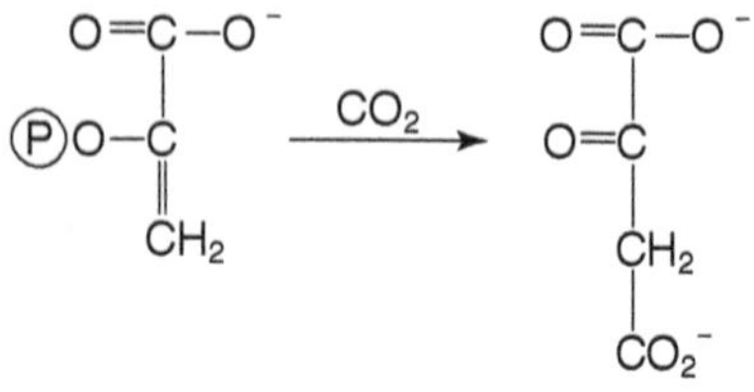

$$ⓅO{-}C(=CH_2){-}C(=O){-}O^- \xrightarrow{CO_2} O{=}C(CH_2CO_2^-){-}C(=O){-}O^-$$

PEP carboxylase has a lower K_m for CO_2 than does Rubisco. Further, O_2 is a very poor substrate for this enzyme. This means that, at relatively low concentrations of CO_2, the delivery of carbon into photosynthesis products is more efficient than in C3 plants and oxygenation doesn't occur.

PEP carboxylase is concentrated in special **mesophyll** cells in the outer part of the leaf. This means that the cells most exposed to the atmosphere are the most efficient at converting CO_2 into organic products. Photosynthesis involving Rubisco is more prominent in the **bundle sheath cells** located in the inner part of the leaf around the veins that carry compounds between different parts of the plant.

After PEP carboxylase makes the oxaloacetate, it is transported to the bundle sheath cells. First, NADPH reduces it to malate, and it is then transported to the bundle sheath cells. In the bundle sheath cells, malic enzyme cleaves the malate to pyruvate and CO_2 for Rubisco. This generates NADPH as well, so the C4 cycle consumes no reducing equivalents. Pyruvate is transported from the bundle sheath back to the mesophyll cells where it is rephosphorylated to phosphoenolpyruvate, expending the equivalent of two ATP "high-energy phosphates."

Overall, the C4 cycle consumes two ATP equivalents to deliver a CO_2 to Rubisco. During active photosynthesis, this is not a problem—plenty of ATP exists from the action of Photosystems I and II. Why, then, don't C4 plants take over the world? Probably because the increased energy demands make these plants less efficient under conditions where sunlight is limited. Consistent with this idea, C4 plants are mostly confined to tropical climates, while the C3 plants predominate in more temperate regions.

CHAPTER 4
NITROGEN FIXATION, ASSIMILATION, AND ELIMINATION

Reduced Nitrogen

Just as photosynthesis or chemosynthesis must reduce carbon dioxide in the atmosphere before it can be used in biological reactions, so must biological nitrogen change from elemental N_2 to a -3 level, as in ammonia, NH_3. In terrestrial and aquatic systems, reduced nitrogen is often a *limiting nutrient* for plant growth. About half of all global nitrogen fixation occurs industrially in a process that requires a metal catalyst at 500 atmospheres pressure and at 300° Celsius. In contrast, biological nitrogen fixation takes place in a much less extreme environment (about 25° and 1 atmosphere pressure) in the roots of leguminous plants or in bacteria, using enzymatic catalysts. Eventually it returns to the atmosphere as N_2. The overall process is referred to as the **nitrogen cycle**.

The Nitrogen Cycle

Reduced nitrogen is used for the synthesis of cellular components. All organisms can incorporate ammonia nitrogen into amino acids, purine and pyrimidine bases, and so forth, so the level of NH_3 is the most useful for cell metabolism. Most plants and bacteria can reduce NO_3^- and NO_2^- to NH_3. Nitrate and nitrite can be formed from more reduced forms of nitrogen through bacterial metabolism. Nitrate and nitrite are also produced atmospherically from elemental nitrogen, especially by burning organic compounds or during the heating of the atmosphere by lightning. Thus, in the nitrogen cycle, nitrogen shuttles through the +5 (NO_3^-), +3 (NO_2^-), 0 (N_2), and -3 (NH_3) levels. See Figure 4-1.

Figure 4-1

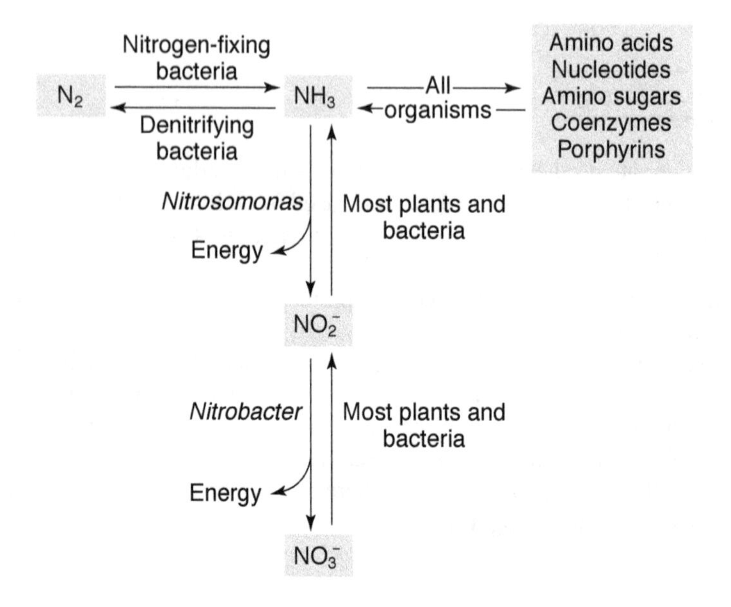

Bacterial Nitrogenase

Nitrogen fixation is done by bacteria. The bacteria that carry out nitrogen fixation occur in either a free-living form or as symbionts living in the roots of leguminous plants, such as soybeans, clover, and peas. Nitrogen fixation involves the use of ATP and reducing equivalents derived from primary metabolism. The overall reaction is catalyzed by **nitrogenase**.

$$8H^+ + N_2 + 8\ e^- + 16\ ATP + 16\ H_2O \rightarrow 2\ NH_3 + H_2 + 16\ ADP + 16\ P_i + 16\ H^+$$

Nitrogenase is a two-protein complex. One component, called **Nitrogenase Reductase (NR)** is an iron-containing protein that

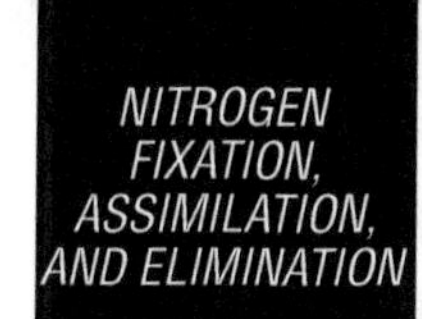

accepts electrons from ferredoxin, a strong reductant, and then delivers them to the other component, called **Nitrogenase**, or **Iron-Molybdenum protein**. See Figure 4-2.

Figure 4-2

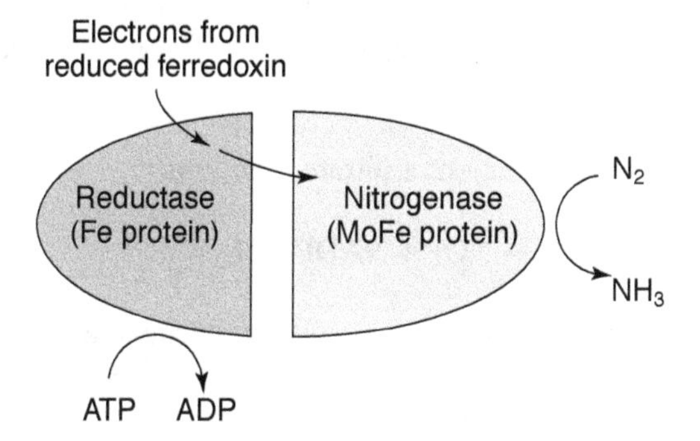

Nitrogenase first accepts electrons from NR and protons from solution. Nitrogenase binds a molecule of molecular nitrogen (releasing H_2 at the same time), and then accepts electrons and protons from NR, adding them to the N_2 molecule, eventually releasing two molecules of ammonia, NH_3. Release of molecular hydrogen, H_2, apparently is an intrinsic part of nitrogen fixation. Many nitrogen-fixing systems contain an enzyme, hydrogenase, which harvests the electrons from molecular hydrogen and transfers them back to ferredoxin, thus saving some of the metabolic energy that is lost during nitrogen reduction.

A major part of the energy of photosynthesis in nodulated plants is used for N_2 fixation. At least sixteen ATPs are hydrolyzed during the reduction of a single nitrogen molecule. This drain of the energy from photosynthesis severely limits the growth of plants that fix nitrogen. For example, the yield of useful energy (protein, carbohydrate, and oil) from a field of maize is much greater than from a field of soybeans.

Nitrogenase is extremely sensitive to oxygen. Root nodules of nitrogen-fixing plants contain an oxygen-binding protein, leghemoglobin, which protects nitrogenase by binding molecular oxygen.

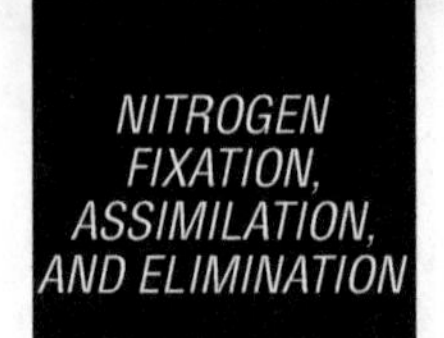

Similar mechanisms operate in the action of nitrate reductase and nitrite reductase. Both of these substances are produced from ammonia by oxidation. Plants and soil bacteria can reduce these compounds to provide ammonia for metabolism. The common agricultural fertilizer ammonium nitrate, NH_4NO_3, provides reduced nitrogen for plant growth directly, and by providing a substrate for nitrate reduction. NADH or NADPH is the electron donor for nitrate reductase, depending on the organism.

The first step is the reduction of nitrate to nitrate.

$$NO_3^- + NADPH + H^+ \rightarrow NO_2^- + NADP + H_2O$$

Step two involves nitrite reductase reducing nitrite to ammonia.

$$NO_2^- + 7\,H^+ + 6\,e^- \rightarrow NH_3 + 2\,H_2O$$

NO^- (nitrite) and NH_2OH (hydroxylamine) are intermediates in the reaction but do not dissociate from nitrite reductase.

Ammonium Utilization

Ammonia is toxic at high concentrations, even though ammonium ion, NH_4^+ is an intermediate in many reactions. For its utilization, ammonia must be incorporated into organic forms, transferred, and then incorporated into other compounds, for example, amino acids and nucleotides. The amino acids **glutamine** and **glutamate** and the compound **carbamoyl phosphate** are the key intermediates of nitrogen assimilation, leading to different classes of compounds.

Glutamate

Glutamate dehydrogenase has a relatively high K_m for ammonia. The high Michaelis constant, K_m, means that this system operates most effectively when ammonia is relatively abundant. Incorporation of ammonia uses **reducing equivalents.**

glutamate dehydrogenase:

$$NH_4^+ + NADPH + {}^-OOC-C(=O)-CH_2-CH_2-COO^- \longrightarrow {}^+H_3N-CH(COO^-)-CH_2-CH_2-COO^- + NADP + H_2O$$

Glutamine
Glutamine synthetase requires ATP energy for ammonia incorporation, using glutamate as an acceptor.

$$NH_3^+ + {}^+H_3N-CH(COO^-)-CH_2-CH_2-C(=O)O^- + ATP \longrightarrow {}^+H_3N-CH(COO^-)-CH_2-CH_2-C(=O)NH_2 + ATP + Pi$$

Glutamate from glutamine
Glutamine can be a precursor for the synthesis of glutamate, with the reaction of **glutamate synthase**, also known as GOGAT (glutamine: 2-oxyglutarate aminotransferase).

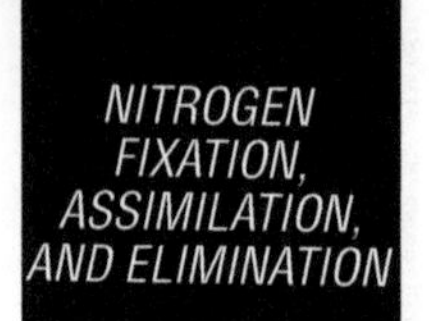

$$^{-}O_2C{-}C(=O){-}CH_2{-}CO^{-}{-}C(=O)O^{-} + H_3\overset{+}{N}{-}CH(CO_2^-){-}CH_2{-}CH_2{-}C(=O)NH_2 \rightleftharpoons H_3\overset{+}{N}{-}CH(CO_2^-){-}CH_2{-}CO_2^- + H_3N{-}C(CO_2^-){-}H{-}CH_2{-}CO_2^- \rightarrow C{-}CH_2{-}CH_2{-}CO_2$$

The preceding reactions indicate that two ways exist of making glutamate:

glutamate dehydrogenase

or

glutamine synthetase + glutamate synthase.

The relative activity of these two pathways depends on the metabolic conditions of the cell. Ammonia is lost from the cell relatively easily. When reduced nitrogen is relatively abundant, glutamate dehydrogenase is more active, because its K_m for ammonia is higher than is that of glutamine synthetase. This may be expected to occur in bacteria growing in a medium containing a high concentration of ammonium salts. On the other hand, when ammonia is relatively nonabundant (for example, when it is supplied by fixation), the pathway involving glutamine synthetase and glutamate synthase is more active. This second pathway uses ATP energy to move ammonia to an amino acid that cannot leak out of the cell.

Further Fates of Incorporated Ammonia

The reduced nitrogen is transferred from glutamate and glutamine into a variety of compounds that participate in a variety of reactions in the cell.

Amino acids. Glutamate (along with aspartate) is a key substrate and product in transamination (aminotransferase) reactions for amino acid interconversions. Aminotransferases carry out the general reaction:

α-amino acid(1) +α-keto acid(2) $\rightleftharpoons$ α-keto acid(1) + α-amino acid(2)

Aminotransferases operate in both directions. Their mechanism uses the cofactor pyridoxal phosphate to form **Schiff bases** with amino groups, as shown in Figure 4-3.

Figure 4-3

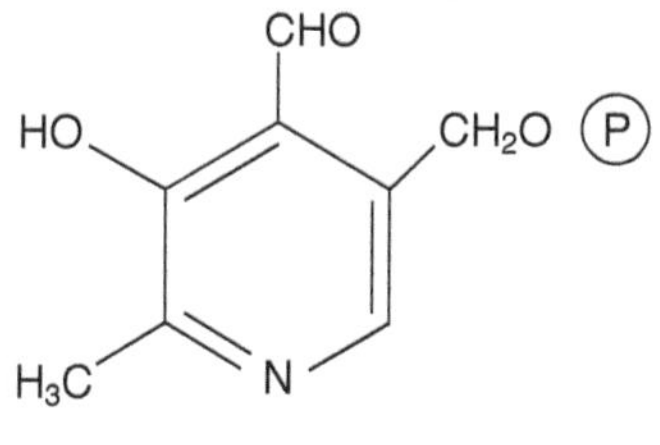

Pyridoxal phosphate

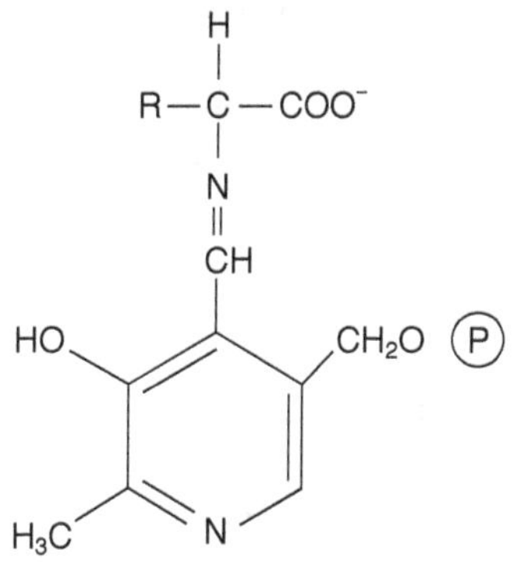

Schiff base between amino acid and pyridoxal phosphate

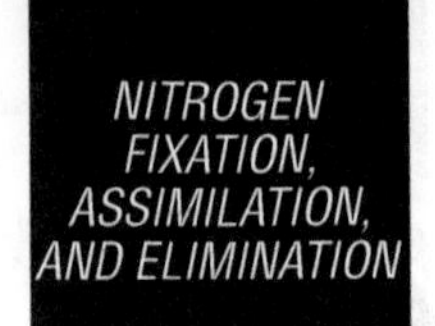

The pyridoxal group is bound to the enzyme by a Schiff base with the ε-amino group of a lysine side chain. This Schiff base is displaced by the amino group of amino acid(1), for example, glutamate. The keto acid, for example, α-ketoglutarate, is released, leaving the amino group on the cofactor, which is now in the **pyridoxamine** form. The rest of the reaction is now the reverse of the first step: The keto group of the second substrate forms a Schiff base with the pyridoxamine, and amino acid(2) is released, with the regeneration of the lysine Schiff base of the enzyme, ready to carry out another cycle.

Nutritionally, humans derive their pyridoxal coenzyme from vitamin B_6. Most symptoms of vitamin B_6 deficiency apparently result from the involvement of the coenzyme in the biosynthesis of neurotransmitters and the niacin group of NAD and NADPH rather than from amino acid deficiency.

Glutamine

The amino group on glutamine is the nitrogen source of a variety of products, including aromatic amino acids, purine and pyrimidine bases, and amino sugars. Glutamine synthetase is therefore an important step in the assimilation of ammonia. Because the enzyme uses ATP, it needs to be regulated to prevent energy wasting. In *bacterial cells,* two enzymes regulate glutamine synthetase. First, the enzyme is subject to **feedback inhibition**. Each of the many end products to which GS serves as a precursor partially inhibits the GS reaction. Feedback inhibition in the living body of a plant or animal depends on the **enzymatic modification** of the GS protein. A separate regulatory system senses the ratio of glutamate to α-ketoglutarate in the cell. If the ratio of these two compounds is high, an enzyme, uridylyl transferase, transfers a UMP group from UTP to a regulatory protein, called P_{II}. The UMP-P_{II} protein associates with another enzyme, adenylyl transferase, and the active adenylyl transferase transfers an AMP from ATP to each of the 12 subunits of glutamine synthetase. This shuts down the enzyme activity

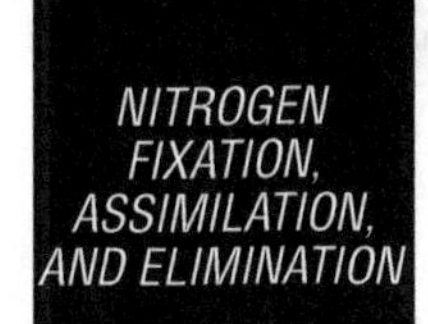

almost completely. Intermediate amounts of adenylation result in intermediate levels of enzyme activity. Thus, the level of nitrogen assimilation is regulated in response to the needs of the bacterial cell.

Carbamoyl phosphate

Carbamoyl phosphate is an "activated ammonia" group that is important in the biosyntheses of the amino acid arginine and of the pyrimidine nucleotides found in DNA and RNA.

Figure 4-4

$$H_2N-C(=O)-O-P(=O)(O^-)-O^-$$

Carbamoyl phosphate

The bacterial carbamoyl phosphate synthetase reaction uses either glutamine or ammonia as substrate.

$NH_3 + HCO_3 + 2ATP \rightarrow$ Carbamoyl phosphate + 2 ADP + P_i

Glutamine + HCO_3 + 2ATP $\rightarrow$ Carbamoyl phosphate + 2 ADP + P_i + Glutamate

In eukaryotic cells, the two enzymes are in different cellular compartments. Form I uses ammonia and is mitochondrial; its function is to provide activated ammonia for arginine biosynthesis (and urea formation during Nitrogen elimination). Form II uses glutamine and is cytoplasmic; it functions in pyrimidine biosynthesis.

Urea

In humans, excess amino groups are detoxified by conversion to urea. Ammonia in high concentrations is toxic to all organisms. For example, humans with kidney failure die because the level of ammonia increases in the bloodstream. Organisms that live in water, including fish and microorganisms, simply eliminate ammonia directly and allow it to be diluted by their environment. Terrestrial organisms, on the other hand, convert the ammonium groups to other compounds that are less toxic. Birds and reptiles synthesize **uric acid** as the primary nitrogen excretion compound. The white crystals of uric acid constitute the unsightly part of bird droppings; the poor solubility of uric acid makes the results hard to remove from a car or a statue in a park, for example. Mammals synthesize urea, ultimately from carbonate and ammonium ions.

Figure 4-5

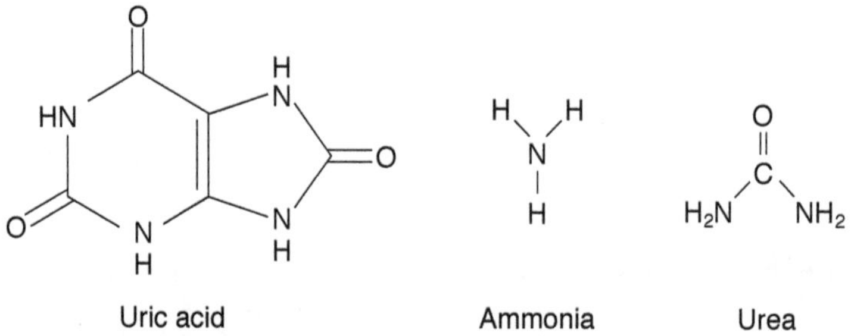

Uric acid — Ammonia — Urea

In mammals, muscle breakdown or excess protein intake results in an imbalance between the fates of the carbon chains and the amino nitrogen. Unlike fat (lipid storage) or glycogen (carbohydrate storage), excess amino acids are not stored in polymeric form for later utilization. The carbon chains of amino acids are generally metabolized into tricarboxylic acid (TCA) cycle intermediates, although it is also possible to make ketone bodies such as acetoacetate from some. Conversion to TCA intermediates is easy to see in some instances. For example, alanine is directly transaminated to pyruvate.

$$H_3\overset{+}{N}-CH(CH_3)-COO^- + O=C(COO^-)-CH_2-CO_2^- \rightleftharpoons CO_2^--C(=O)-CH_3 + H_3\overset{+}{N}\,CH(CO_2)-CH_2-CO_2$$

More complex processes convert other amino acids, especially the branched chained ones.

The nutritional consequences of an excess protein diet are the same as those of an excess carbohydrate or excess fat diet: lipid biosynthesis and fat deposition. Additionally, the protein amino groups must be detoxified and eliminated. The nutritional consequences of a diet lacking complete protein—that is, one that doesn't supply the essential amino acids in the proportions needed for synthesis of proteins and neurotransmitters—also include excess ammonia generation. In this case, muscle proteins are degraded to supply enough of the **limiting essential** amino acid. The other amino acids are broken down, with the carbon chains metabolized into carbohydrates (and, potentially lipid). The leftover amino groups must then be eliminated as urea.

Biochemistry of the urea cycle

First the ammonium ions must be activated through conversion to carbamoyl phosphate by the **mitochondrial form** of carbamoyl phosphate synthetase.

$NH_4^+ + HCO_3^- + 2\ ATP \rightarrow$ Carbamoyl Phosphate $+ 2\ APP + P_i$

$$CO_2 + NH_4^+ + 2ATP \longrightarrow O=C(NH_2^+)-O\text{-}ⓟ + 2ADP + Pi$$

The entry of activated ammonia into the urea cycle occurs by the ornithine transcarbamoylase reaction where the carbamoyl group is transferred to the side chain amino group of the non-protein amino acid, **ornithine**. Ornithine has five carbons; its carbon chain therefore has the same length as that of arginine. The product of the ornithine transcarbamoylase reaction is the amino acid **citrulline**.

Figure 4-6

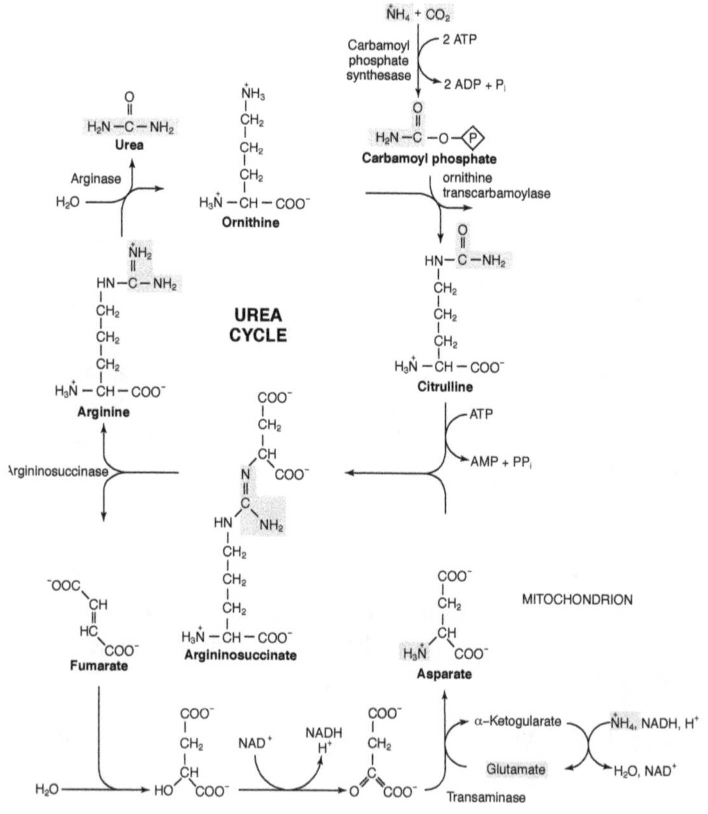

$NH_4^+ + CO_2$
Carbamoyl phosphate synthesase
2 ATP
2 ADP + P_i
Carbamoyl phosphate
ornithine transcarbamoylase
Citrulline
ATP
AMP + PP_i
Urea
Arginase
H_2O
Ornithine
UREA CYCLE
Arginine
Argininosuccinase
Argininosuccinate
Fumarate
Asparate
MITOCHONDRION
H_2O
NAD^+
NADH H^+
α-Ketogularate
Glutamate
Transaminase
NH_4^+, NADH, H^+
H_2O, NAD^+

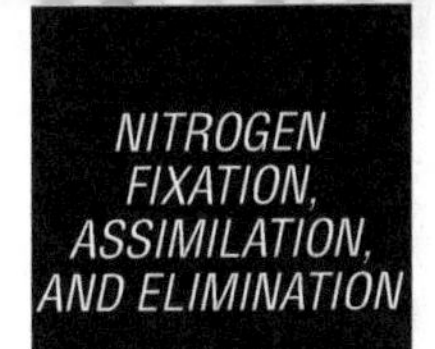

The addition of the second ammonia to the backbone involves the joining of the carbamoyl group of citrulline to the χ-amino group of aspartate, leading to a complex compound, arginosuccinate. ATP is hydrolyzed in this step.

$$\mathrm{H_2N{-}C(=O){-}NH{-}(CH_2)_3{-}HC(NH_3^{\oplus}){-}COO^-} \;+\; \mathrm{{}^-OOC{-}CH_2{-}HC(NH_3^+){-}COO^-} \xrightarrow[\text{AMP} + 2\mathrm{P_i}]{\text{ATP}} \mathrm{{}^-OOC{-}CH_2{-}HC(COO^-){-}NH{-}C(=NH_2^{\oplus}){-}NH{-}(CH_2)_3{-}HC(NH_3^{\oplus}){-}COO^-}$$

The next step is cleavage of arginosuccinate to arginine and fumarate by the enzyme arginosuccinate lyase. Lyases cleave bonds with the creation of a double bond in one of the products. In this case, the double bond is the carbon-carbon double bond of fumarate.

$$\mathrm{{}^-OOC{-}CH_2{-}HC{-}NH{-}C(=NH_2^{\oplus}){-}NH{-}(CH_2)_3{-}HC(NH_3^{\oplus}){-}COO^-} \longrightarrow \mathrm{H_2N{-}C(=NH_2^{\oplus}){-}NH{-}(CH_2)_3{-}HC(NH_3^{\oplus}){-}COO^-} \;+\; \mathrm{{}^-OOC{-}CH{=}HC{-}COO^-}$$

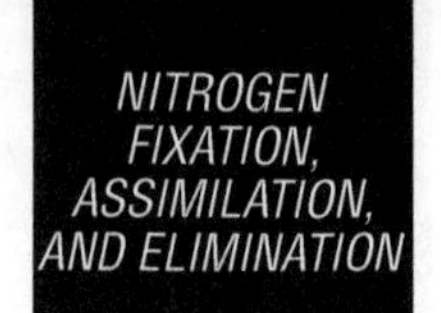

The final step is the release of urea by the enzyme arginase, which regenerates ornithine.

$$H_2N{-}C({=}\overset{\oplus}{N}H_2){-}NH{-}(CH_2)_3{-}CH(\overset{\oplus}{N}H_3){-}COO^- \longrightarrow \overset{\oplus}{N}H_3{-}(CH_2)_3{-}CH(\overset{\oplus}{N}H_3){-}COO^- + H_2N{-}C({=}O){-}NH_2$$

The other part of the urea cycle that has occurred is the conversion of the carbons of aspartate to fumarate. The fumarate is recycled back to oxaloacetate through TCA cycle reactions in the mitochondrion. Transamination with glutamate regenerates aspartate. The glutamate comes from the glutamate dehydrogenase reaction.

$$^-OOC{-}CH{=}CH{-}COO^- \xrightarrow{H_2O} {}^-OOC{-}CH(OH){-}CH_2{-}COO^- \xrightarrow[NADH + H^+]{NAD} {}^-OOC{-}C({=}O){-}CH_2{-}COO^- \xrightarrow[\text{Keto acid}]{\text{Amino acid}} {}^-OOC{-}CH(\overset{+}{H_3N}){-}CH_2{-}COO^-$$

The urea is excreted through the kidneys and broken down to carbon dioxide and ammonia by plants and microorganisms. The enzyme **urease** causes this conversion.

CHAPTER 5
AMINO ACID METABOLISM: CARBON

Principles of Amino Acid Metabolism

Some catabolic reactions of amino acid carbon chains are easy transformations to and from TCA cycle intermediates—for example, the transamination of alanine to pyruvate. Reactions involving 1-carbon units, branched-chain, and aromatic amino acids are more complicated. This chapter starts with "1-carbon metabolism" and then considers the catabolic and biosynthetic reactions of a few of the longer side chains. Amino acid metabolic pathways can present a bewildering amount of material to memorize. Perhaps fortunately, most of the more complicated pathways lie beyond the scope of an introductory course or a review such as this. Instead of a detailed listing of pathways, this chapter concentrates on general principles of amino acid metabolism, especially those that occur in more than one pathway.

1-Carbon Metabolism

The 1-carbon transformations require two cofactors especially: folic acid and vitamin B_{12}.

Folic acid

Several compounds that interfere with folic acid metabolism are used in clinical medicine as inhibitors of cancer cells or bacterial growth.

Folic acid participates in the activation of single carbons and in the oxidation and reduction of single carbons. Folate-dependent single-carbon reactions are important in amino acid metabolism and in biosynthetic pathways leading to DNA, RNA, membrane lipids, and neurotransmitters.

Folic acid is a composite molecule, being made up of three parts: a ***pteridine*** ring system (6-methylpterin), ***para-aminobenzoic acid***, and ***glutamic acid***. The glutamic acid doesn't participate in the coenzyme functions of folic acid. Instead, folic acid in the interior of the cell may contain a "chain" of three to eight (1–6) glutamic acids, which serves as a negatively charged "handle" to keep the coenzyme inside cells and/or bound to the appropriate enzymes. The pteridine portion of the coenzyme and the *p*-aminobenzoic acid portion participate directly in the metabolic reactions of folate.

To carry out the transfer of 1-carbon units, NADPH must reduce folic acid two times in the cell. The "rightmost" pyrazine ring of the 6-methylpterin is reduced at each of the two N-C double bonds. See Figure 5-1.

Figure 5-1

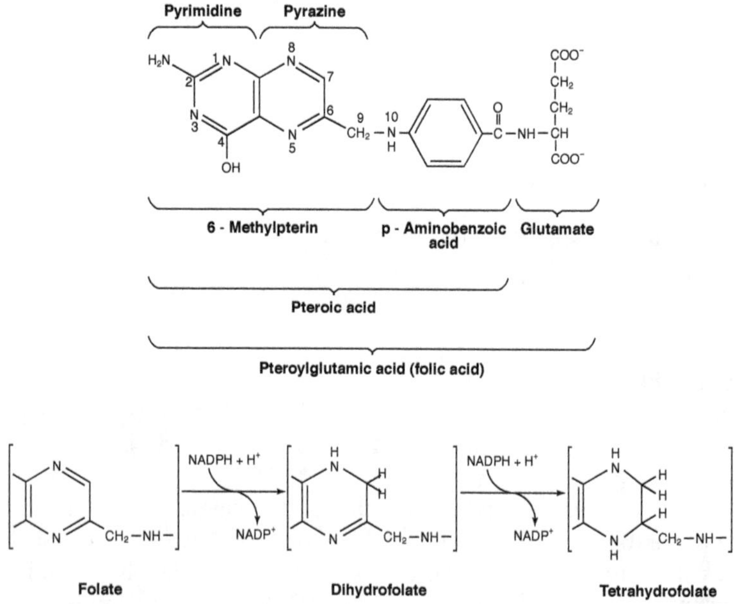

The resulting 5,6,7,8-tetrahydrofolate is the acceptor of 1-carbon groups.

Tetrahydrofolate accepts methyl groups, usually from serine. The product, N^5,N^{10}-methylene-tetrahydrofolate, is the central compound in 1-carbon metabolism. Tetrahydrofolate can also accept a methyl group from the complete breakdown of glycine. In Figure 5-2, only the N^5, C^6, and N^{10} atoms of the pteroic acid are shown for clarity.

Figure 5-2

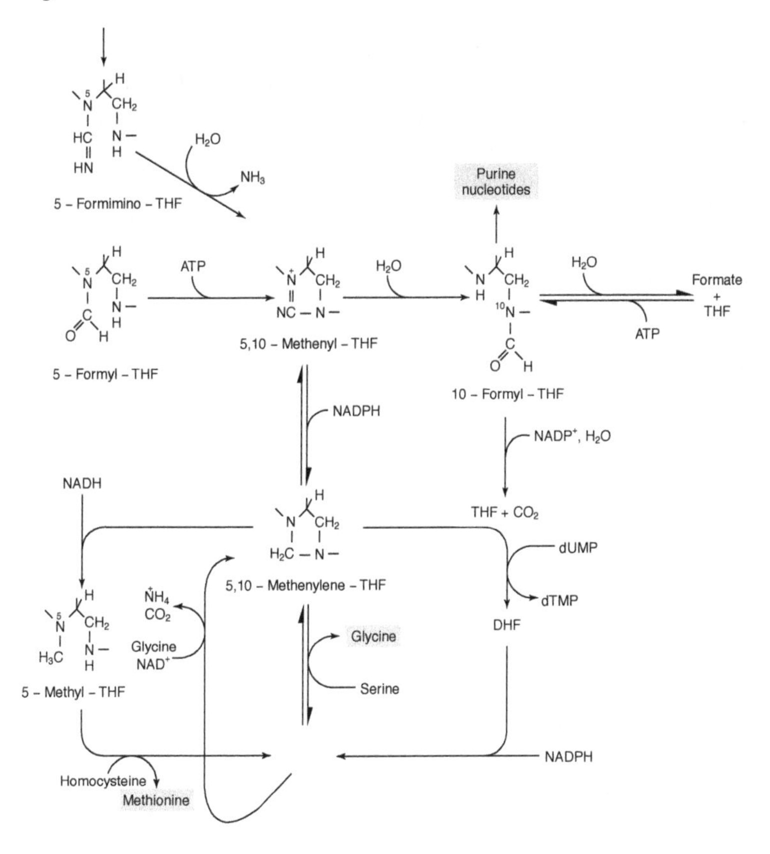

The N5,N10-methylene-tetrahydrofolate can either donate its single-carbon group directly, be oxidized by NADP to the *methenyl* form, or be reduced by NADH to the *methyl* form. Depending on the biosynthetic pathway involved, any of these species can donate the 1-carbon group to an acceptor. The methylene form donates its methyl group during the biosynthesis of thymidine nucleotides for DNA synthesis, the methenyl form donates its group as a formyl group during purine biosynthesis, and the methyl form is the donor of the methyl group to sulfur during methionine formation.

Folate antagonists as antimicrobial drugs

Sulfanilamide is the simplest of the **sulfa drugs**, used as antibacterial agents. Note the similarity of sulfanilamide to *p*-aminobenzoic acid as shown in Figure 5-3. Because its shape is similar to that of *p*-aminobenzoic acid, sulfanilamide inhibits the growth of bacteria by interfering with their ability to use *p*-aminobenzoic acid to synthesize folic acid. Sulfa drugs were the first **antimetabolites** to be used in the treatment of infectious disease. Because humans don't make folic acid, sulfanilamide is not toxic to humans in the doses that inhibit bacteria. This ability to inhibit bacteria while sparing humans made them useful in preventing or treating various infections.

Figure 5-3

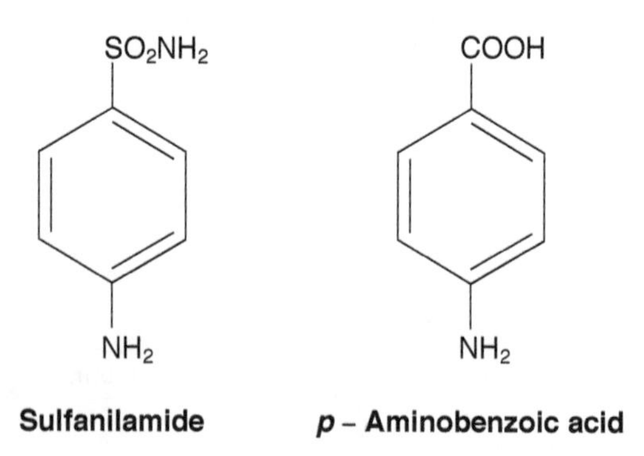

Vitamin B_{12}

In humans, vitamin B_{12} participates in two reactions only, but they are essential to life. Humans who cannot absorb vitamin B_{12} die of pernicious anemia if untreated (now by injection of the vitamin; formerly by eating large amounts of raw liver). Vitamin B_{12} contains a *cobalt* metal ion bound to a *porphyrin* ring. Cobalt normally forms six coordinate bonds. Besides the four bonds to the nitrogens of the porphyrin, one bond is to a ring nitrogen of dimethylbenzamidine. The final bond is to a cyanide ion in the vitamin, or to the 5' carbon of adenosine in the active coenzyme.

Vitamin B_{12} is essential for the methylmalonyl-CoA mutase reaction. Methylmalonyl-CoA mutase is required during the degradation of odd-chain fatty acids and of branched-chain amino acids. Odd-chained fatty acids lead to propionyl-CoA as the last step of β-oxidation. Methylmalonyl-CoA can be derived from propionyl-CoA by a carboxylase reaction similar to that of fatty acid biosynthesis. The cofactor for this carboxylation reaction is biotin, just as for acetyl-CoA carboxylase. The reaction of methylmalonyl-CoA mutase uses a free radical intermediate to insert the methyl group *into* the dicarboxylic acid chain. The product is succinyl-CoA, a Krebs cycle intermediate. The catabolisms of branched-chain lipids and of the branched-chain amino acids also require the methylmalonyl-CoA mutase, because these pathways also generate propionyl-CoA.

Vitamin B_{12} activates methyl groups for *methionine biosynthesis* by binding them to the Co ion at the sixth position. The methyl group donor to B_{12} is 5-methyl tetrahydrofolate. The methyl-B_{12} donates its methyl group to *homocysteine,* forming *methionine.*

Figure 5-4

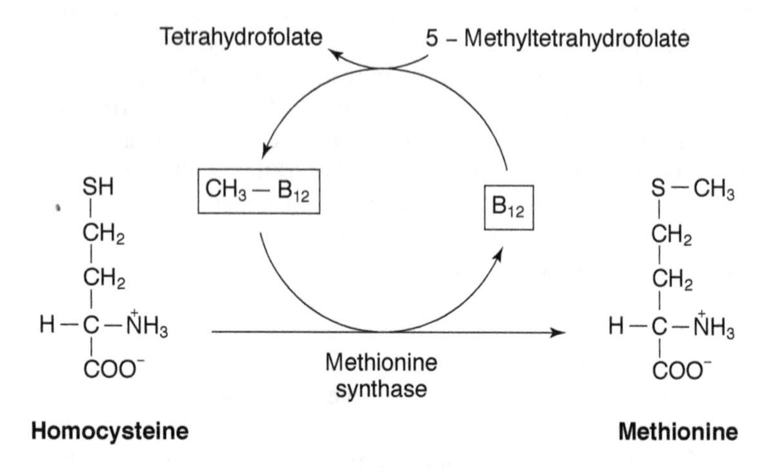

Besides being incorporated into proteins, methionine is the source of methyl groups for several important reactions, including the modification of cellular RNAs and the biosynthesis of lipids.

Methyltransferases

Most methyltransferase reactions use *methionine* as the source of methyl groups. The actual methyl donor is *S-adenosyl methionine,* abbreviated S-AdoMet, or more colloquially, SAM. S-Adenosylmethionine is made from methionine and ATP. Note how the reaction consumes all three "high-energy" phosphate bonds of the ATP as shown in Figure 5-5.

Figure 5-5

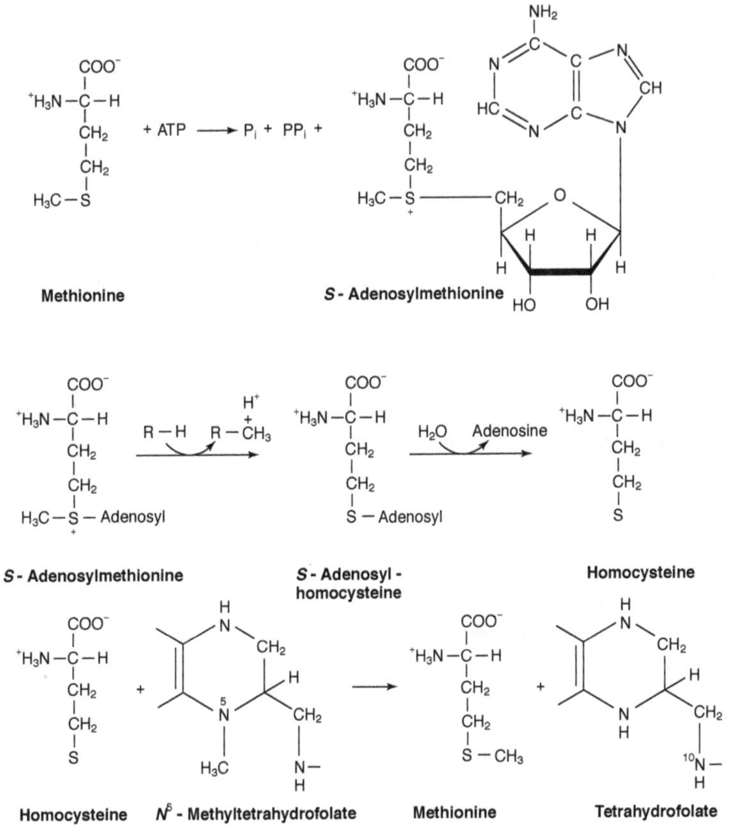

Methionine + ATP → P_i + PP_i + *S*-Adenosylmethionine

S-Adenosylmethionine + R−H → $R-CH_3$ + H^+ + *S*-Adenosyl-homocysteine

S-Adenosyl-homocysteine + H_2O → Adenosine + Homocysteine

Homocysteine + N^5-Methyltetrahydrofolate → Methionine + Tetrahydrofolate

Methyl transfer from S-AdoMet is highly favored chemically and metabolically. First, transfer of the methyl group relieves a positive charge on the Sulfur of S-AdoMet. Secondly, the bond between the Sulfur and the 5' carbon of the adenosine is rapidly hydrolyzed, leaving ***homocysteine*** and free ***adenosine***. This last step is important, because the product remaining when S-AdoMet gives up its methyl group, ***S-adenosyl-homocysteine***, is a potent *inhibitor* of methyltransferases. Cleavage of this product removes the inhibitor from the reaction.

Homocysteine itself is converted to methionine by the transfer of a methyl group from *N5-methyl-tetrahydrofolate* to homocysteine, regenerating methionine. The methyl group of N5-methyl-tetrahydrofolate is derived from serine, originally, so the net effect of this pathway is to move methyl groups from serine to a variety of acceptors, including homocysteine, nucleic acid bases, membrane lipids, and protein side chains. Serine itself is easily made from 3-phosphoglycerate by an amino transfer reaction, followed by cleavage of the phosphate.

Homocysteine isn't harmless. Evidence from population studies indicates that high levels of homocysteine in the blood are correlated with heart disease. Folic acid supplements may prevent this problem by ensuring that the homocysteine is rapidly converted to methionine. Similarly, pregnant women generally take folic acid supplements to prevent their babies from being born with neural tube defects. The mechanism for its action isn't known, but the folic acid may help decrease the level of homocysteine in this case as well.

Amino Acid Biosynthesis

The metabolism of the carbon chains of amino acids is varied. In humans and laboratory rats, half of the twenty amino acids found in proteins are *essential* and must be supplied in the diet, either from plant, animal, or microbial sources. The other half can be made from other compounds, especially from the products of carbohydrate metabolism. You can remember the essential amino acids by using a mnemonic:

Very Many Hairy Little Pigs Live In The Torrid Argentine

which translates to:

Valine Methionine Histidine Leucine Phenylalanine Lysine
Isoleucine Threonine Tryptophan Arginine

Only human babies require Arginine (which is made in the urea cycle) and histidine.

Nonessential amino acids

Many of the nonessential amino acids are derived directly from intermediate products of the TCA cycle or glycolysis. The key reactions include:

Transamination, as in the synthesis of alanine from pyruvate.

$$\underset{CH_3}{\overset{CO_2^-}{C{=}O}} + H_3\overset{+}{N}-\underset{R}{\overset{CO_2^-}{CH}} \rightleftharpoons H_3\overset{+}{N}-\underset{CH_3}{\overset{CO_2^-}{C}}-H + \underset{R}{\overset{CO_2^-}{C{=}O}}$$

Amidation via amidotransferases, as in glutamine synthetase:

$$H_3\overset{+}{N}-\overset{CO_2^-}{C}(CH_2CH_2COO^-)-H + NH_3^+ + ATP \longrightarrow H_3\overset{+}{N}-\overset{CO_2^-}{CH}-CH_2-CH_2-C(=O)-NH_2 + ADP + Pi$$

A more complex series of reactions synthesizes other nonessential amino acids. For example, the carbon chain of cysteine derives from serine and the sulfur derives from homocysteine (which results after methyl donation from S-AdoMet).

Essential amino acids

For the most part, plants and free-living microorganisms make these. The fact that humans can't synthesize them except by degrading protein has practical consequences. Any single amino acid can be limiting in the diet. Thus, for example, an individual who does not get enough tryptophan in the diet can't compensate by eating more methionine. The only way to get enough tryptophan would be to break down muscle to release enough tryptophan to support life. This means that a person can literally starve to death while getting what appears to be enough calories and protein. In the 1970s, a fad of eating only hydrolyzed liquid collagen and vitamins in an attempt to lose weight appeared. These commercial products were removed from the market after several people died of heart problems. The cause was simple: Collagen is very poor in aromatic amino acids. People were breaking down their muscle, including heart muscle, to provide these amino acids for the synthesis of essential proteins. Eventually they died of cardiac insufficiency. Animal and dairy proteins usually contain a balanced supply of amino acids. That is, they are complete proteins. Plant proteins aren't always complete, so individuals who eat primarily vegetable sources of proteins must eat complementary foods to supply a full set of amino acids. This is the reason for the coexistence of rice and beans in so many cuisines. The combination supplies all the essential amino acids, while neither food does so on its own.

Biosynthesis of aromatic amino acids

The aromatic amino acids, phenylalanine, tryptophan, and tyrosine, are all made from a common intermediate: **chorismic acid**. Chorismic acid is made by the condensation of erythrose-4-phosphate and phosphoenol pyruvate, followed by dephosphorylation and ring closure, dehydration and reduction to give **shikimic acid**. Shikimic acid is phosphorylated by ATP and condenses with another phosphoenol pyruvate and is then dephosphorylated to give chorismic acid.

Figure 5-6

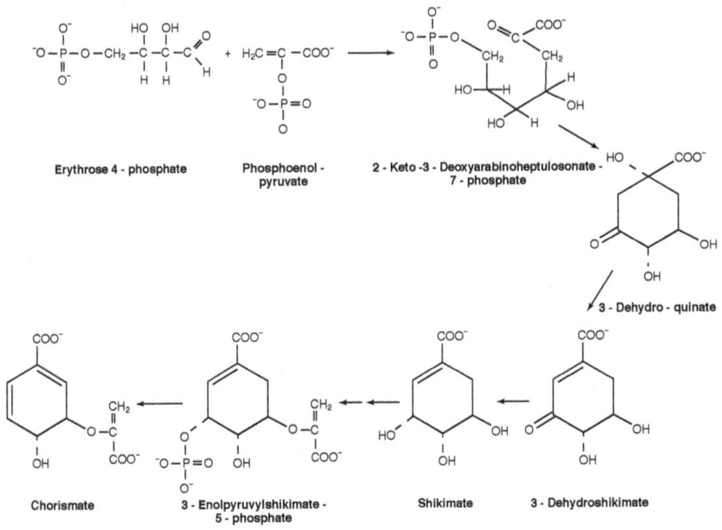

Erythrose 4 - phosphate

Phosphoenol - pyruvate

2 - Keto -3 - Deoxyarabinoheptulosonate - 7 - phosphate

3 - Dehydro - quinate

3 - Dehydroshikimate

Shikimate

3 - Enolpyruvylshikimate - 5 - phosphate

Chorismate

The condensation of phospho-shikimic acid with phosphoenol pyruvate is catalyzed by the enzyme 3-enoylpyruvoylhikimate-5-phosphate synthase, or **EPSP synthase** for short. This reaction is specifically inhibited by the herbicide **glyphosate**.

Glyphosate is toxic to plants and free-living microorganisms because it inhibits aromatic amino acid biosynthesis. On the other hand, it is extremely nontoxic to humans and animals because humans derive their amino acids from the diet. Additionally, it is broken down in the soil, so it is non-persistent. The only problem with glyphosate herbicides is that they will kill crop plants as readily as weeds. Recently, genetically engineered crop varieties have been introduced which are resistant to the herbicide, allowing weeds to be killed preferentially.

Once chorismate is produced, it can be converted to either tryptophan, tyrosine, or phenylalanine by distinct pathways. Additionally,

many organisms can hydroxylate phenylalanine to tyrosine. This reaction explains why tyrosine isn't an essential amino acid in humans, even though humans can't make any aromatic amino acids from simple precursors.

Pathway determination by biochemical genetics

The use of genetic mutants determined the complex pathways that lead to the amino acids. A mutant is an organism that has a different DNA sequence from its parent(s). Mutant bacteria that require a specific compound for growth are called **auxotrophs**. The first step in pathway determination is to assemble a large collection of auxotrophic mutants that can't make the compound of interest.

The pathway for the biosynthesis of tryptophan illustrates the overall process. First, the auxotrophic mutants had to be organized into classes that correspond to the biochemical steps in the pathway. Mutant strains were examined for their Ability to excrete compounds that allowed the growth of other mutants. These *cross-feeding* experiments were carried out for pairs of mutants. Thus, the compound excreted from one group of mutants, called Group A, was able to support the growth of other mutants, including Group E. On the other hand, Group E mutants could not excrete a compound that allowed Group A mutants to grow. The fact that Group A mutants excreted a product that allowed Group E mutants to grow, but not vice versa, means that the compound excreted by Group A mutants occurs *after* the compound excreted by Group E mutants (if any). This is a fundamental idea in determining biochemical pathways.

To understand this idea, consider a biochemical pathway to be like a series of entrances to a highway. Normally, one can enter the highway at any on-ramp and get to one's destination. But what happens if a wreck occurs between entrances A and E? The wreck blocks the traffic, and the unlucky drivers who got on the highway ahead of the wreck (those who got on at E) can't get to their destination. On the other hand, those who have gotten on at A remain free to travel and won't even notice a delay. See Figure 5-7.

Figure 5-7

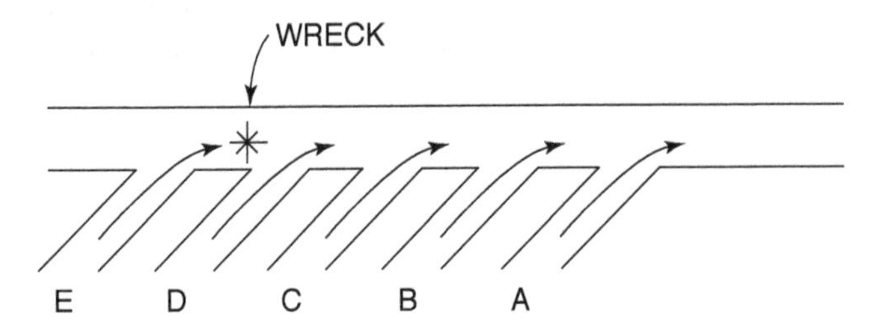

The same logic was used to identify the groups of mutants who allowed the growth of other mutants, or who were able to grow on the compounds excreted by other groups of mutants. In the case of tryptophan, five such **complementation groups** existed.

Allowed the growth of :

Group	*A*	*B*	*C*	*D*	*E*
A	No	Yes	Yes	Yes	Yes
B	No	No	Yes	Yes	Yes
C	No	No	No	Yes	Yes
D	No	No	No	No	Yes
E	No	No	No	No	No

From these data, it should be fairly easy to deduce that complementation group E comes last in the pathway while complementation group A comes first.

This did not establish the pathway completely. The next step was to identify the compounds that are excreted and allow growth of the other mutants. Next, the pathway was established biochemically by identifying and purifying the individual proteins that carry out the steps corresponding to each complementation group, showing that the enzymes behaved kinetically in the way that the other analyses predicted was necessary. These studies showed that the proteins encoded by Groups A and B are part of a complex (called tryptophan synthase). Similarly, the gene products of the D and E groups associate to form the first enzyme, anthranilate synthetase. The overall pathway is as follows:

Figure 5-8

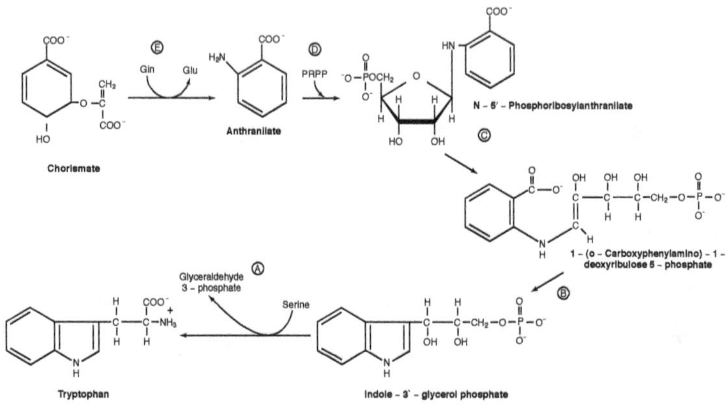

Feedback inhibition

You can see that biosynthesis of amino acids is a complex, energy-requiring process. Almost all biosynthetic pathways are regulated so they don't produce too much of the product. This process, **feedback inhibition,** *occurs at the first committed step of the biochemical pathway.* In tryptophan biosynthesis, the first committed step is the synthesis of anthranilic acid from chorismate. When excess tryptophan is available, either through biosynthetic reactions or from the environment, it is able to bind to a site on anthranilate synthetase. This binding to the allosteric regulatory site causes the enzyme to be less active. This means that chorismate is diverted to the tyrosine and phenylalanine pathways. Higher levels of tryptophan can partially inhibit the synthesis of chorismate, thus further sparing the cell's energy reserves.

Amino acid catabolism

Excess amino acids are degraded, rather than stored, by almost all biological systems. Seed formation in plants and the synthesis of yolk and proteins in eggs constitute the major exceptions. Thus, a "high-protein" diet normally provides little benefit. Most healthy individuals need a relatively small amount of dietary protein, unless they are growing children. A typical "Western diet," with a large meat intake, isn't necessary for health.

The products of amino acid breakdown are of two kinds. Ketone bodies—that is, acetoacetate and hydroxybutyrate—are formed from the catabolism of the branched-chain amino acids, lysine and some aromatic amino acids. Tricarboxylic acid (TCA) cycle intermediates, including pyruvate and glutamate, are formed from most of the other aliphatic and aromatic amino acids. Amino acids whose metabolisms produce ketone bodies such as acetoacetate are called **ketogenic**; amino acids whose metabolisms produce TCA cycle intermediates are called **glucogenic**, because TCA cycle intermediates are substrates for gluconeogenesis. Individual amino acids can be exclusively ketogenic, exclusively glucogenic, or both. Only leucine and lysine are considered to be exclusively ketogenic, and some suspicion remains that they may also give rise to TCA cycle intermediates.

Amino acids are also **precursors** to biologically important compounds. In animals, removal of the carboxylic acid group from amino acids creates many neurotransmitters.

Dopamine, which is essential for the control of movement, is made from tyrosine (or phenylalanine, because tyrosine is a direct product of phenylalanine metabolism).

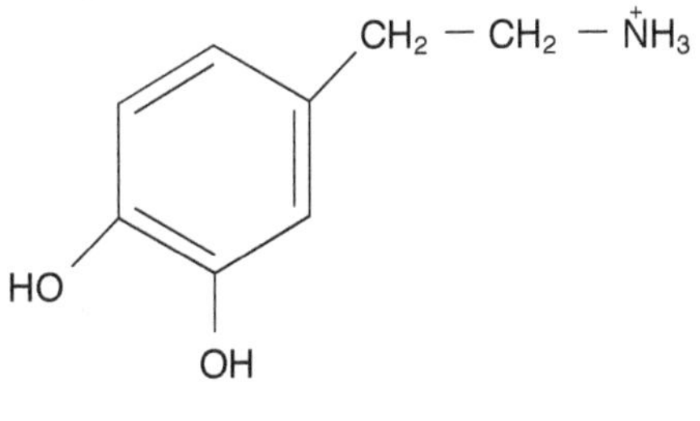

$CH_2 - CH_2 - \overset{+}{N}H_3$

HO

OH

Serotonin, a "sleep-inducing" transmitter, is made from tryptophan.

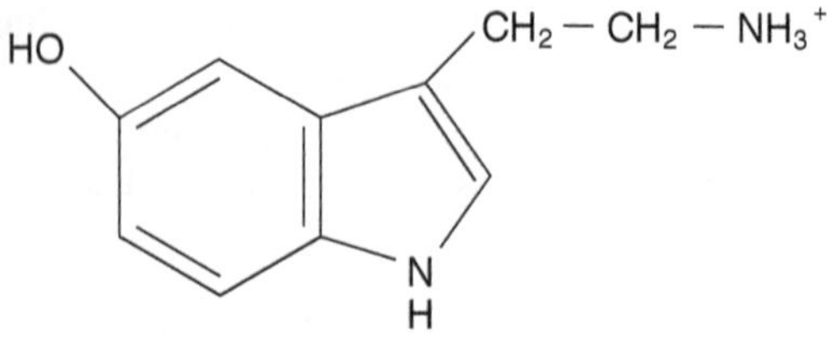

HO
$CH_2 - CH_2 - NH_3^+$
N
H

γ-aminobutyric acid (GABA) is made from glutamate.

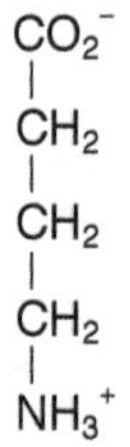

CO_2^-
|
CH_2
|
CH_2
|
CH_2
|
NH_3^+

Inborn errors of amino acid metabolism

An inability to degrade amino acids causes many genetic diseases in humans. These diseases include phenylketonuria (PKU), which results from an inability to convert phenylalanine to tyrosine. The phenylalanine is instead transaminated to phenylpyruvic acid, which is excreted in the urine, although not fast enough to prevent harm. PKU was formerly a major cause of severe mental retardation. Now, however, public health laboratories screen the urine of every newborn child in the United States for the presence of phenylpyruvate, and place children with the genetic disease on a synthetic low-phenylalanine diet to prevent neurological damage.

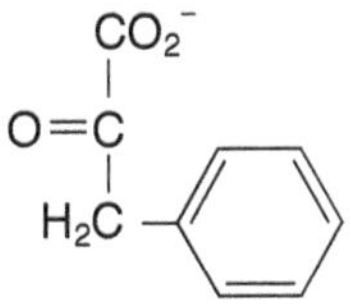

Heme biosynthesis

Hemoglobin and myoglobin carry blood oxygen. Heme is iron bound to a **porphyrin** ring. Porphyrin biosynthesis begins with δ-aminolevulinic acid, the condensation product of the amino acid glycine with succinyl-CoA. The ring is formed by the "head to tail" condensation of two δ-aminolevulinic acid molecules to form porphobilinogen.

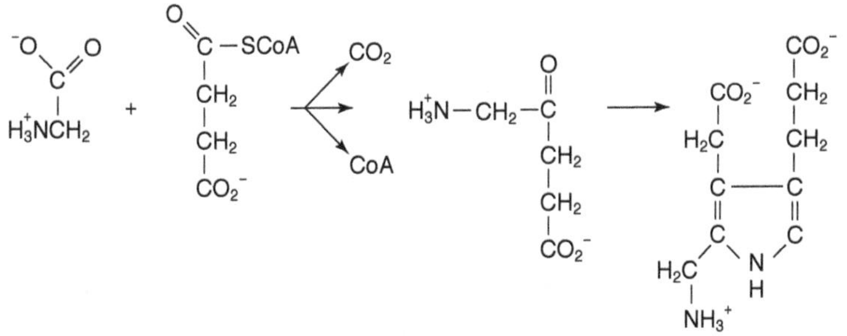

Note the Schiff base formation that forms the 5-membered ring.

Heme contains four of these rings. Four porphobilinogens condense head to tail to form the first tetrapyrrole species, which is then circularized to form the porphyrin skeleton. Further modifications, followed by Fe(II) addition, lead to heme.

Figure 5-9

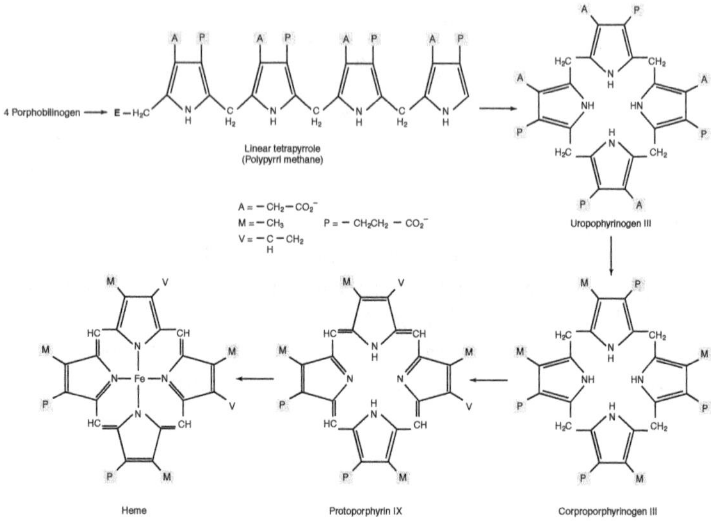

4 Porphobilinogen → E–H_2C

Linear tetrapyrrole
(Polypyrri methane)

A = $-CH_2-CO_2^-$
M = $-CH_3$
V = $-\underset{H}{C}-CH_2$
P = $-CH_2CH_2-CO_2^-$

Uropophyrinogen III

Heme

Protoporphyrin IX

Corproporphyrinogen III

Inborn metabolic diseases that interfere with heme biosynthesis are called **porphyrias.** Porphyrias have a variety of symptoms. A deficiency in the enzyme responsible for the condensation of porphobilinogen to the 4-membered ring system leads to a condition called **acute intermittent porphyria,** which is characterized by occasional episodes of abdominal pain and psychiatric symptoms. Defects in the later enzymes of the pathway lead to an excess accumulation of the uroporphobilinogens in the tissues, where they cause a variety of symptoms, including hairy skin, skeletal abnormalities, light sensitivity, and red urine. Individuals with this disease are still anemic—a condition that can be alleviated somewhat by the heme acquired from drinking blood. This combination of traits sounds like the werewolf and vampire legends of Europe, which may have their base in this rare biochemical disease.

CHAPTER 6
PURINES AND PYRIMIDINES

Roles of nucleotides

Purine and pyrimidine nucleotides fill a variety of metabolic roles. They are the "energy currency" of the cell. In some cases, they are signaling molecules, acting like hormones directly or as transducers of the information. They provide the monomers for genetic information in DNA and RNA.

Purine and Pyrimidine Structures

The pyrimidine bases have a 6-membered ring with two nitrogens and four carbons.

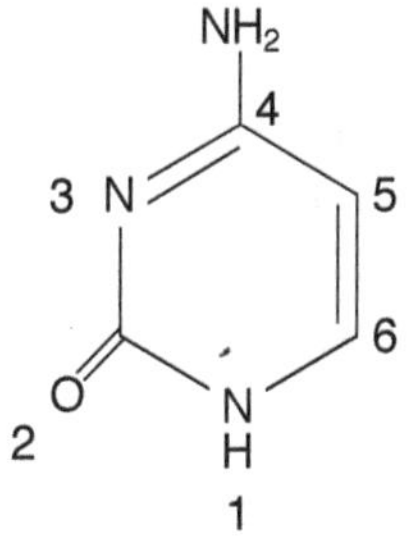

Cytosine

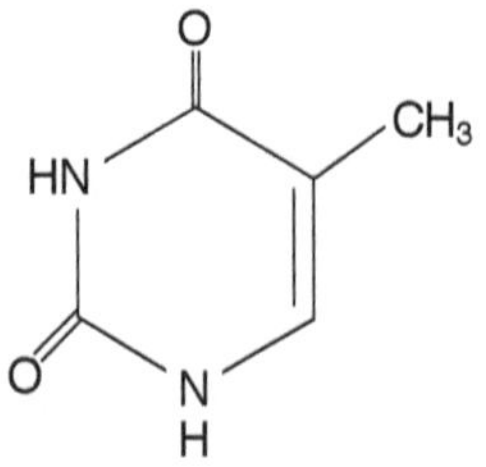

Thymine

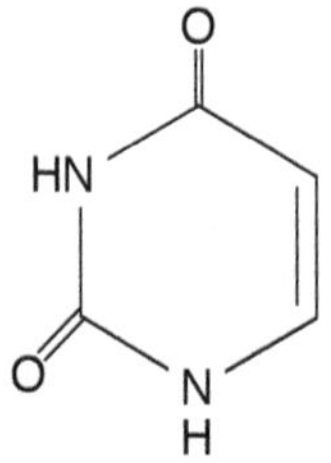

Uracil

The purine bases have a 9-membered double-ring system with four nitrogens and five carbons.

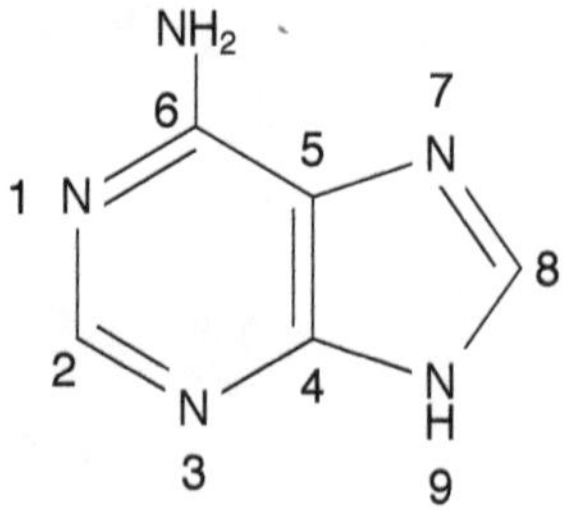

Adenine

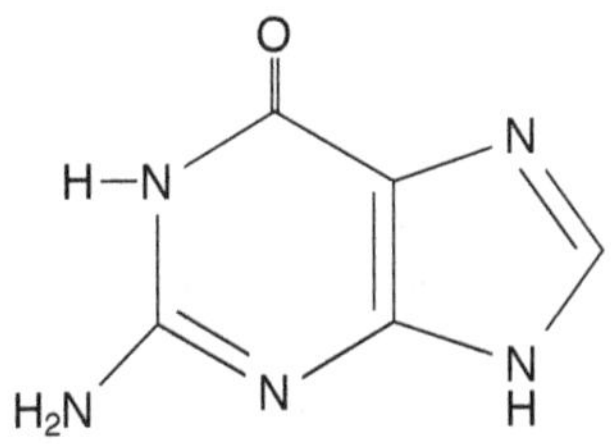

Guanine

Although both purine and pyrimidine rings have one 6-membered component with two nitrogens and four carbons, the purines and pyrimidnes are not related metabolically. Distinct pathways for purine biosynthesis and degradation and for pyrimidine biosynthesis and degradation, exist in all organisms.

The combination of a 5-membered carbohydrate ring and a purine or pyrimidine is called a **nucleoside**. The rings are numbered as shown in the following figure. The two rings of a nucleoside or nucleotide must be distinguished from each other, so the positions of the sugar carbons are denoted with a ' (prime) notation. If one or more phosphates exist on the carbohydrate, the combination is called a **nucleotide**. For example, ATP is a nucleotide.

Deoxyguanosine is a nucleoside as is 2'-O-methyladenosine.

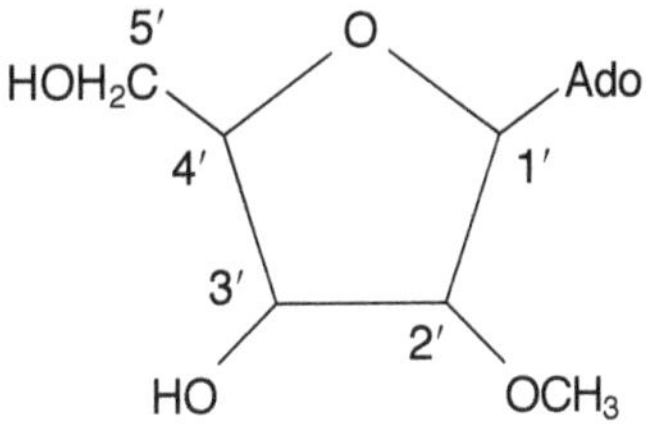

3'-cytidine monophosphate is a nucleotide.

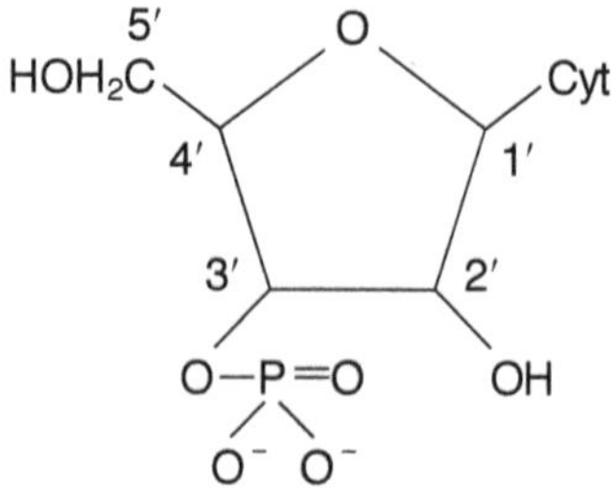

Salvage and Biosynthetic Pathways

Nucleotides and nucleosides can be supplied to an organism by either a salvage reaction or by synthesis from smaller precursors. Salvage reactions convert free purine and pyrimidine bases into nucleotides. Additionally, free purines and pyrimidines can be degraded, the purines to the oxidized ring compound uric acid and the pyrimidines to smaller compounds (β-amino acids, not the α-amino acids found in proteins). Finally, purines and pyrimidines can be synthesized from smaller precursors (**de novo synthesis**). Thus three interacting pathways for nucleotides, nucleosides, and the free bases exist: salvage, degradation, and biosynthesis.

This complexity is due to the central role of nucleotides as energy currency, signaling molecules, and precursors to informational macromolecules in the cell. If the supply of nucleotides becomes limiting, cells couldn't make DNA or RNA, for example. Likewise, cells need to have a *balanced* supply of nucleotides, because A and T, as well as C and G, occur at the same proportions in DNA and in similar amounts in RNA. Thus the cell must ensure the availability of an adequate supply of precursors. On the other hand, more ATP is needed in energy storage relative to the other nucleoside triphosphates. Finally, the purine bases themselves and the purine nucleosides are toxic to humans (for a variety of reasons), so they must be readily eliminated.

Salvage pathways

The nucleotide and nucleosides of a cell are continually in flux. For example, DNA and RNA chains are being synthesized in the cell. Even though the overall DNA content of a cell is constant, small stretches are continually being repaired. Part of the repair process is the breakdown of one strand of the DNA double helix into nucleotides, nucleosides, and free bases. Free purines and pyrimidines are converted back into nucleoside triphosphate monomers to be reincorporated into DNA. A common step in this pathway is the reaction of free bases with **phosphoribosyl pyrophosphate (PRPP)** to yield nucleotides. PRPP is a **general activator** of nitrogen ring compounds. For example, PRPP is added to anthranilate during the biosynthesis of tryptophan in bacteria, as mentioned in the chapter on amino acid metabolism. PRPP is made by the activation of ribose-5-phosphate. Ribose-5-phosphate can be made through the pentose phosphate pathway. Apparently, two enzymes exist in all systems—one for purines and one for pyrimidines. The synthesis of the glycosidic bond uses the 1'-pyrophosphate of PRPP as an energy source, and either enzyme transfers the free base to the 1' position of the ribose, making a nucleotide. See Figure 6-1.

Figure 6-1

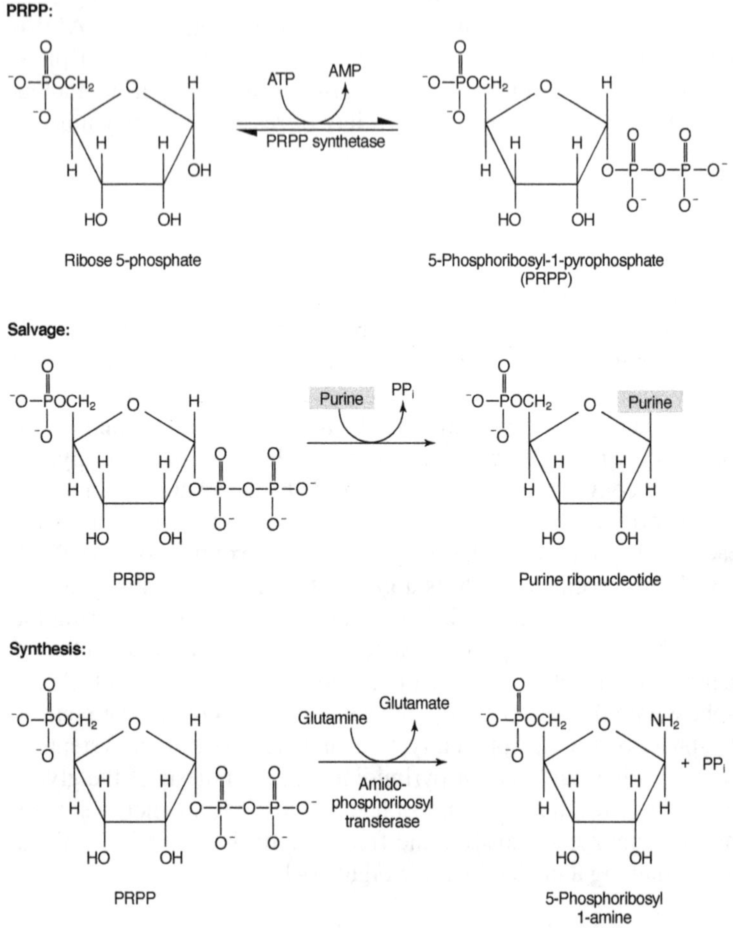

PRPP:

ATP → AMP, PRPP synthetase

Ribose 5-phosphate

5-Phosphoribosyl-1-pyrophosphate (PRPP)

Salvage:

Purine → PP_i

PRPP

Purine ribonucleotide

Synthesis:

Glutamine → Glutamate, Amido-phosphoribosyl transferase

PRPP

5-Phosphoribosyl 1-amine + PP_i

One enzyme uses either guanine or hypoxanthine (adenine with the amino group replaced by an OH). A second enzyme uses free adenine. A third enzyme is specific for uracil and thymine. All the enzymes carry out the same reaction: transfer of the free base to the ribose-5'-monophosphate of PRPP, forming a nucleoside-5'-monophosphate (NMP).

Purine biosynthesis
Purine synthesis uses a PRPP "handle" where the ring is assembled to make a 5' NMP, **inosine monophosphate (IMP)**.

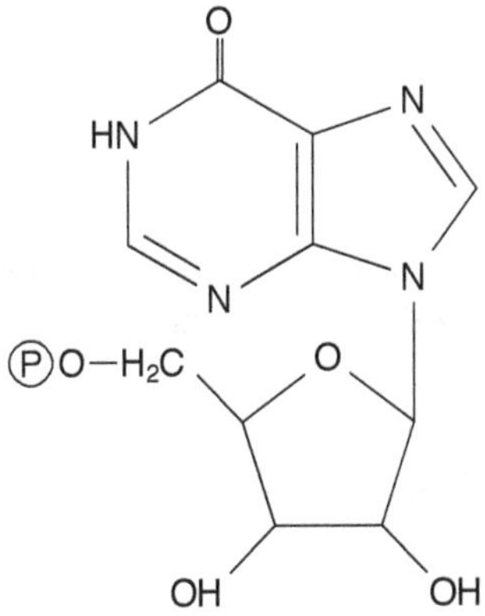

IMP

IMP is the common intermediate in purine biosynthesis, and can be converted to GMP or AMP as needed.

The first reaction in purine biosynthesis is the transfer of the amide from glutamine to PRPP with release of pyrophosphate. The product is **phosphoribosylamine (PRA)**.

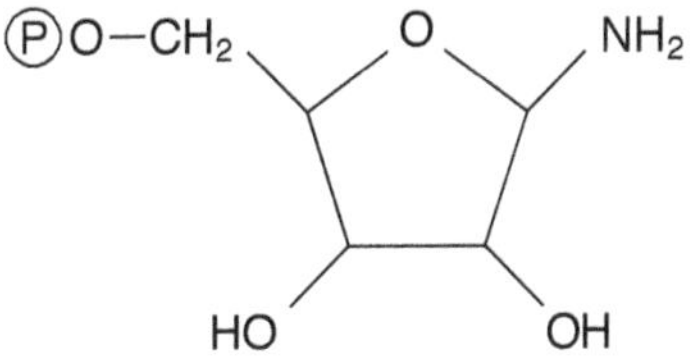

Then the amino acid glycine is transferred to PRA, making glycinamide mononucleotide.

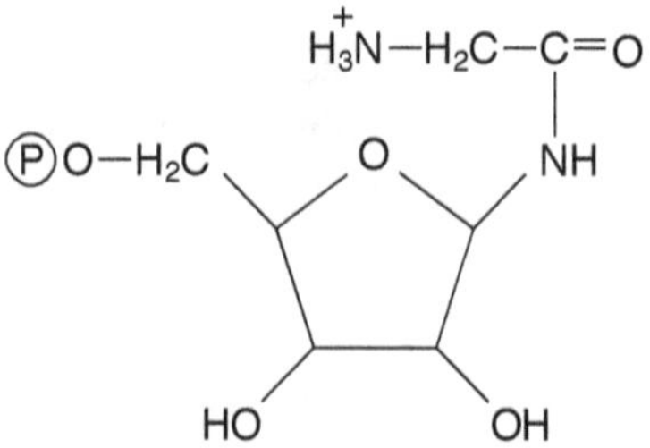

$H_3\overset{+}{N}-H_2C-C=O$

NH

Ⓟ$O-H_2C$ O

HO OH

The amino group of glycine is formylated, with the formyl group being donated by N^{10}-formyl-tetrahydrofolate.

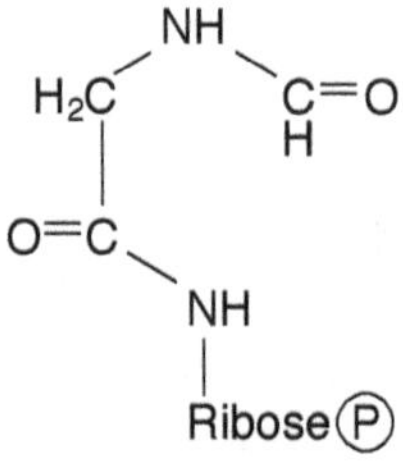

NH

H_2C $C=O$

H

$O=C$

NH

RiboseⓅ

Now the amino NH_2 is transferred to the carboxyl carbon of glycine from glutamin, with ATP as an energy source. This compound, formylglycineamidine ribonucleotide, closes to make the "smaller" (imidazole) ring of the purine. Again, ring closure uses ATP energy.

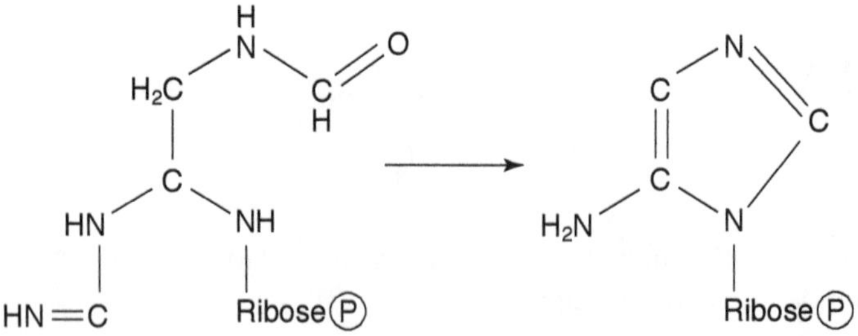

H

N O

H_2C C N

H C

C C

HN NH H_2N N

$HN=C$ RiboseⓅ RiboseⓅ

Now the larger ring is built on the smaller one. A carboxylation reaction with CO_2 starts synthesis of the 6-membered ring.

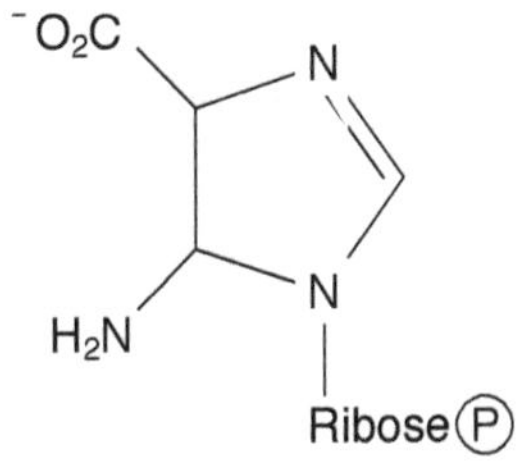

Then the amino group of aspartate is transferred to the carboxyl, making an amide. This condensation uses ATP and the amide is cleaved to release fumarate, leaving behind the imidazole with a 5-amino group (left from the amidation of glycine four steps earlier) and a 4-carboxamide. (Note how this reaction is similar to the formation of arginine during the urea cycle.)

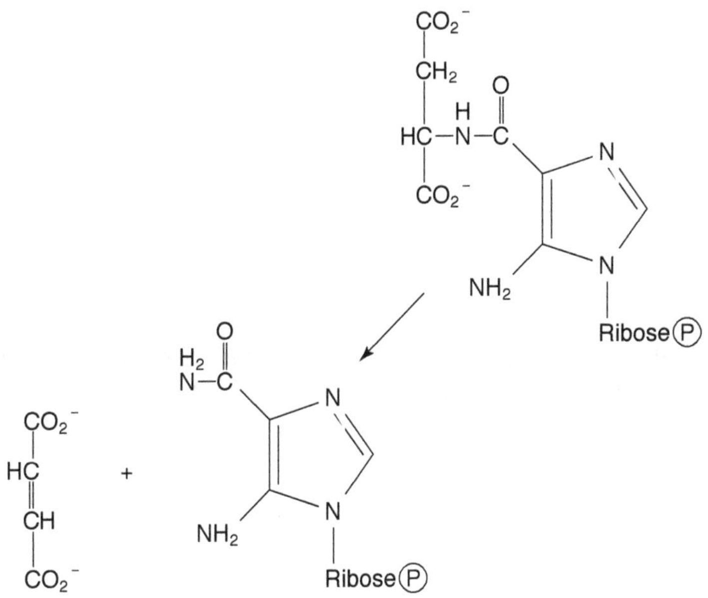

Eight of the nine components of the ring are now present. The last ring component comes from a 1-carbon transfer of a formyl group from N^{10}-formyltetrahydrofolate.

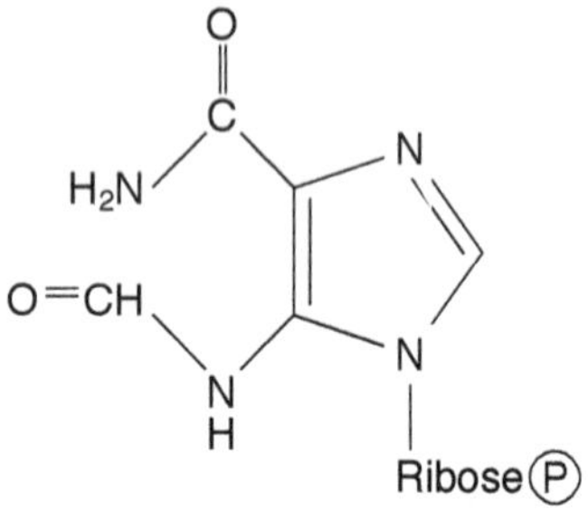

Finally, the ring is closed by dehydration to yield IMP.

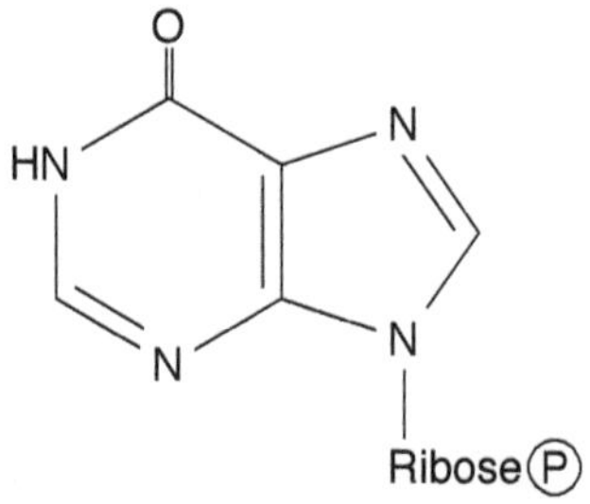

IMP is the key intermediate of purine nucleotide biosynthesis. IMP can react along two pathways that yield either GMP or AMP. Oxidation of the 2 position makes **xanthine monophosphate**, which is transamidated to GMP. Alternatively, the α-amino group of aspartate can replace the ring oxygen of IMP to make AMP. (Note again how this reaction is similar to the synthesis of arginine from citrulline.)

Figure 6-2

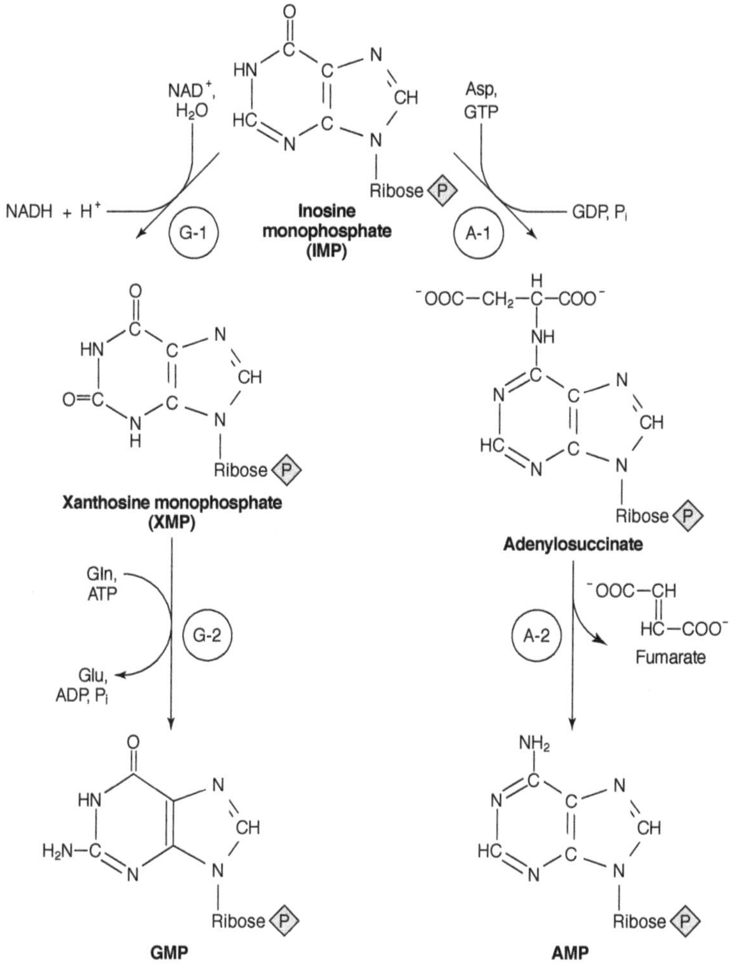

The rates of these two complementary reactions can control the amount of either AMP or GMP present in the cell. Each of these reactions is feedback-inhibited by its nucleotide product. Thus, if more adenosine nucleotides exist than guanosine nucleotides, the synthesis of AMP slows down until the purine nucleotides balance.

Degradation of purine nucleotides

Extra purines in the diet must be eliminated. In mammals, the product of purine breakdown is a weak acid, **uric acid**, which is a purine with oxygen at each of three carbons.

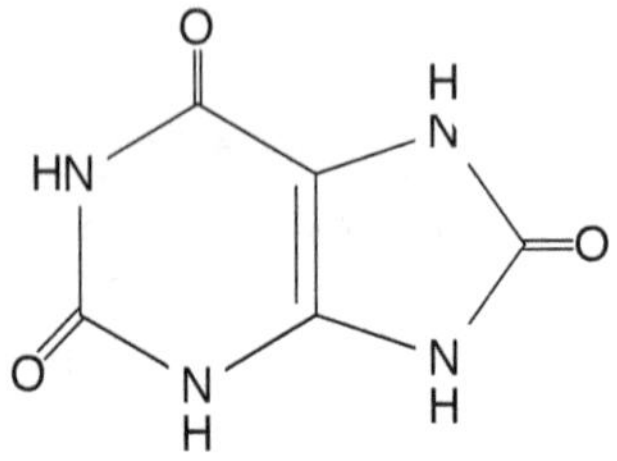

Uric acid is the major nitrogen excretion product in birds and reptiles, where it is responsible for the white, chalky appearance of these droppings. Uric acid is poorly soluble in water, and in humans, formation of uric acid crystals is responsible for the painful symptoms of **gout**. These crystals are deposited in joints (recall that the classic symptom of gout is an inflamed toe).

Adenosine is degraded in a two-step reaction. First, the enzyme **adenosine deaminase** acts on AMP or adenosine nucleoside to yield IMP or inosine.

IMP is cleaved by **phosphorolysis** of the nucleoside to yield hypoxanthine and ribose-1-phosphate. (This reaction is similar to the phosphorolysis of glycogen by glycogen phosphorylase.)

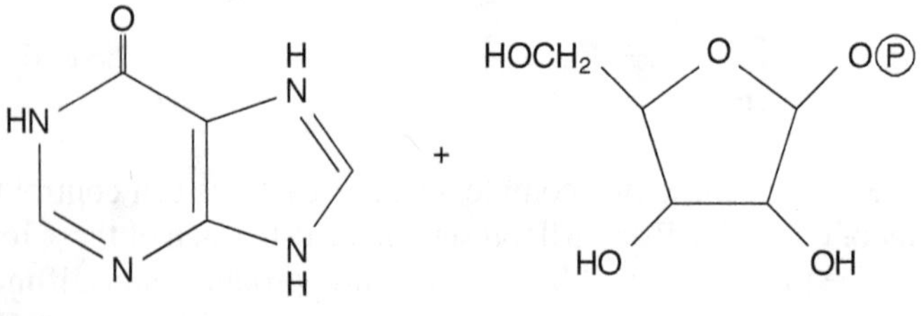

Guanosine is degraded in a two-step reaction sequence. First, guanosine phosphorylase phosphorolyses the nucleoside to free guanine and ribose-1-phosphate.

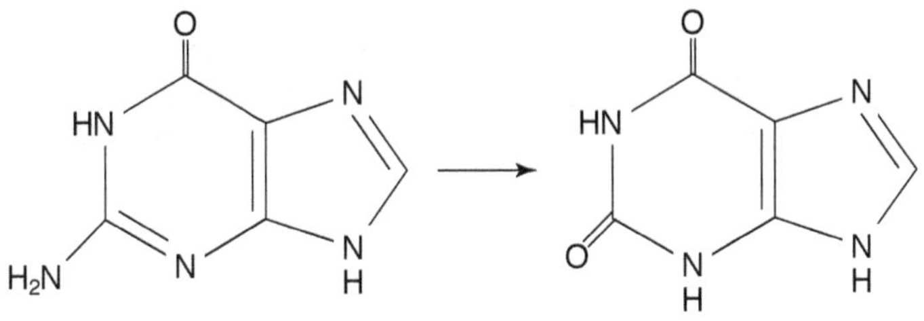

The next reaction is the deamination of guanosine to xanthine. Xanthine needs only one more oxygen to form uric acid.

Xanthine oxidase oxidizes hypoxanthine and xanthine to uric acid, using molecular oxygen, O_2.

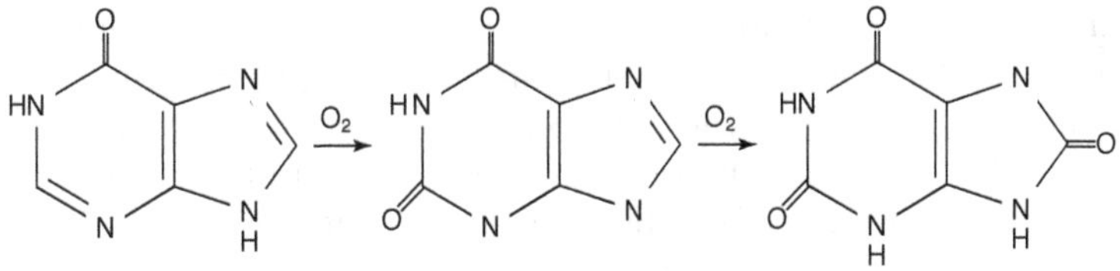

As mentioned earlier, uric acid is only slightly soluble and individuals with impaired secretion or excess production of uric acid are subject to the pain of gout as uric acid precipitates in the joints. Most cases of gout are probably due to impaired excretion of uric acid because of poor kidney function. Because the concentration of uric acid in the blood is near the solubility limit, only a slight impairment of elimination can push the concentration high enough to precipitate uric acid. More frequently nowadays, gout appears in persons whose kidney function is impaired with age, although it is also found in individuals with genetic deficiencies in the level of hypoxanthine-guanine phosphoribosyl transferase. In the latter case, the salvage pathway does not function well, and more purines must be eliminated through their conversion to uric acid.

The drug **allopurinol**, which is an inhibitor of xanthine oxidase, effectively treats gout. Allopurinol is structurally similar to hypoxanthine, except that the 5-membered ring has the positions of the carbon and nitrogens reversed.

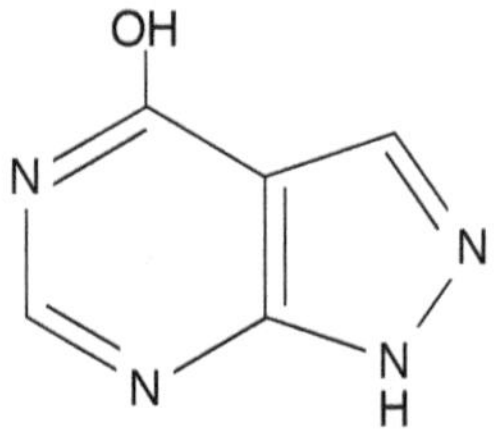

Xanthine oxidase is able to bind allopurinol and catalyze one oxidation, converting it to a compound that is similar to xanthine. However, after that conversion, the enzyme is trapped in an inactive oxidation state and can't carry out its normal function of forming uric acid. Additionally, allopurinol inhibits the de novo (new, from other compounds; not recycled) synthesis of purines, further decreasing the amount of uric acid formed in the blood.

Pyrimidine Metabolism

Although both pyrimidines and purines are components in nucleic acids, they are made in different ways. Likewise, the products of pyrimidine degradation are more water-soluble than are the products of purine degradation.

Pyrimidine biosynthesis

Unlike in purine biosynthesis, the pyrimidine ring is synthesized before it is conjugated to PRPP. The first reaction is the conjugation of carbamoyl phosphate and aspartate to make N-carbamoylaspartate. The carbamoyl phosphate synthetase used in pyrimidine biosynthesis is located in the cytoplasm, in contrast to the carbamoyl phosphate used in urea synthesis, which is made in the mitochondrion. The enzyme that carries out the reaction is aspartate transcarbamoylase, an enzyme that is closely regulated.

Figure 6-3

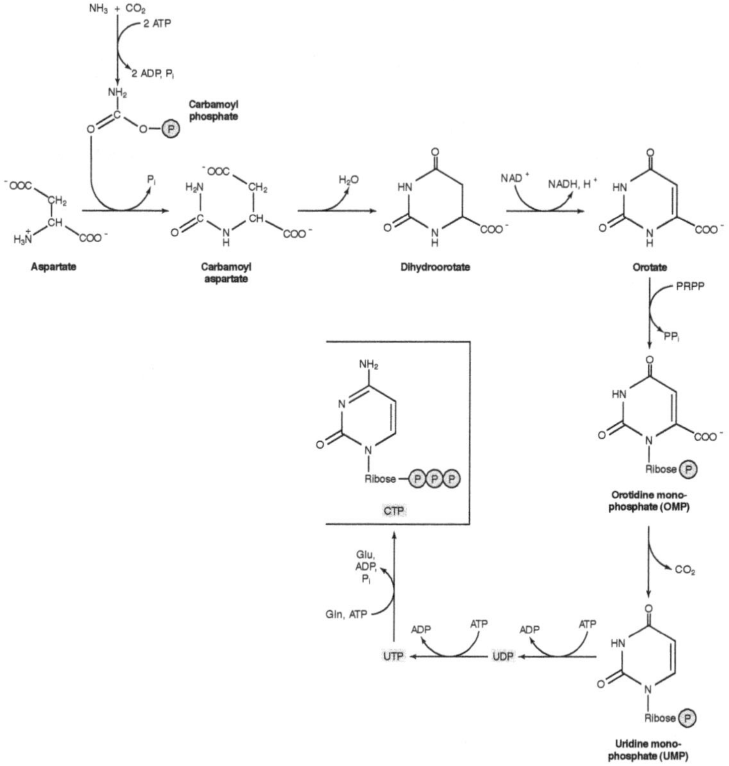

The second reaction is ring closure to form **dihydroorotic acid** by the enzyme dihydroorotase. This circular product contains a 6-membered ring with nitrogen and carbons located in the same positions as in the mature pyrimidine ring.

The third reaction is the oxidation of the ring to form a carbon-carbon bond. The reducing equivalents are transferred to a flavin cofactor of the enzyme dihydroorotate dehydrogenase. The product is **orotic acid**.

Fourth, the orotate ring is transferred to phosphoribosyl pyrophosphate (PRPP) to form a 5' ribose-phosphate, **orotidylic acid**.

Finally orotidylate is decarboxylated to yield UMP, which of course contains one of the bases of RNA. Cellular kinases convert UMP to UTP. Transfer of an amido nitrogen from glutamine by CTP synthetase converts UTP to CTP; this reaction uses an ATP high-energy phosphate.

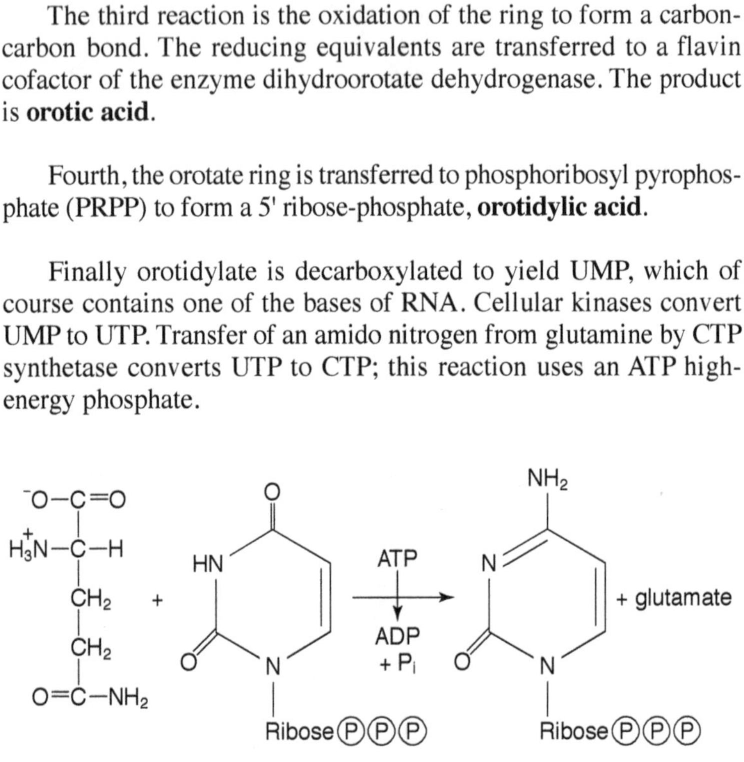

Control
Pyrimidine synthesis is controlled at the first committed step. ATP stimulates the aspartate transcarbamoylase reaction, while CTP inhibits it. CTP is a feedback inhibitor of the pathway, and ATP is a **feed-forward** activator. This regulation ensures that a balanced supply of purines and pyrimidines exists for RNA and synthesis.

Eukaryotic organisms contain a multifunctional enzyme with carbamoylphosphate synthetase, aspartate transcarbamoylase, and dihydroorotase activities. Two mechanisms control this enzyme. First, control at the level of enzyme synthesis exists; the transcription of the gene for the enzyme is reduced if an excess of pyrimidines is present. Secondly, control exists at the level of feedback inhibition by pyrimidine nucleotides. This enzyme is also an example of the phenomenon of **metabolic channeling**: aspartate, ammonia, and carbon dioxide enter the enzyme and come out as orotic acid.

Deoxynucleotide Synthesis

The enzyme DNA polymerase, which uses deoxynucleoside triphosphates as substrates, makes DNA. To ensure enough precursors for DNA synthesis, two reactions must occur. First, the 2' position of the ribose ring of ribonucleotides must be reduced from a C-OH to a C-H before the nucleotides can be used for DNA synthesis. Secondly, the thymine ring must be made by addition of a methyl group to uridine.

Ribonucleotide reductase uses ribonucleoside diphosphates (ADP, GDP, CDP, and UDP) as substrates and reduces the 2' position of ribose.

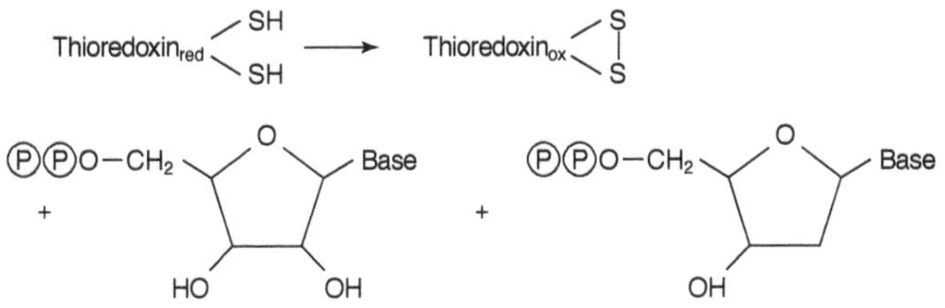

Thioredoxin$_{red}$ (SH, SH) ⟶ Thioredoxin$_{ox}$ (S–S)

+ ⓅⓅO–CH$_2$ (ribose: O, Base, HO, OH) ⟶ + ⓅⓅO–CH$_2$ (deoxyribose: O, Base, OH)

The small protein thioredoxin supplies reducing equivalents to ribonucleotide reductase for the ribose ring reduction. Thioredoxin is itself reduced by another protein, thioredoxin reductase, a flavoprotein. Reduced glutathione can also carry reducing equivalents to ribonucleotide reductase. In both cases, the ultimate source of reducing equivalents is NADPH.

The regulation of ribonucleotide reductase is complex, with many feedback reactions used to keep the supplies of deoxynucleotides in balance. For example, dGTP and dTTP are feedback inhibitors of their own formation. Each is also an activator of the synthesis of the complementary nucleotide (dCDP or dADP), while dATP is an inhibitor of the reductions to make dADP, dCDP, dGDP, and dUDP. These control functions keep the supply of deoxynucleotides in *balance*, so that a roughly equivalent amount of each remains available for DNA synthesis.

Thymidylate synthase

DNA contains thymidine, while RNA contains uridine. The formation of thymidine must be controlled and, more crucially, the formation of dUTP and its incorporation into DNA must be prevented. (Deoxyuridine in DNA can lead to mutations in the sequence and possible genetic defects.)

The thymidylate synthase reaction involves the methylation of deoxy-UMP to deoxy-TMP (thymidylate). Deoxy-UMP is the result of dephosphorylation of the product of the ribonucleotide reductase reaction, dUDP. The conversion of the diphosphate nucleotide to the monophosphate nucleotides helps channel deoxyuridine to thymidylate synthase rather than directly to DNA. N^5,N^{10}-methylene tetrahydrofolate donates the methyl.

Figure 6-4

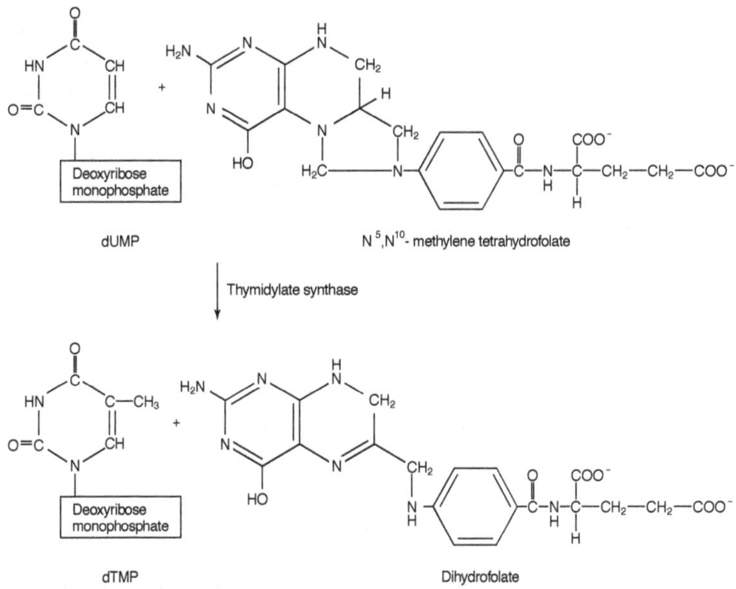

Deoxyribose monophosphate

dUMP

N^5,N^{10}- methylene tetrahydrofolate

Thymidylate synthase

Deoxyribose monophosphate

dTMP

Dihydrofolate

The donation of the methyl group from N^5,N^{10}-methylene tetrahydrofolate leads to the oxidation of the cofactor to dihydrofolate. This points to the importance of dihydrofolate reductase (DHFR) in the functioning of thymidylate synthase. Thus, synthesis of TMP requires a supply of both methyl groups—for example, from serine—and reducing equivalents.

Chemotherapy

Both dihydrofolate reductase and thymidylate synthase reactions are targets for anticancer chemotherapy. Cancer is basically a disease of uncontrolled cell replication, and an essential part of cell replication is DNA synthesis. This means that a requirement exists for deoxynucleotide synthesis for growth. Inhibition of deoxynucleotide synthesis should inhibit the growth of cancer cells.

The compound 5-fluorouridine targets thymidylate synthase. After a nucleoside kinase phosphorylates it, resembles the natural substrate for the enzyme, except that it contains a fluorine where dUMP has a hydrogen. The fluorine isn't removed from the ring by thymidylate synthase, and this causes the ring to remain covalently bound to the enzyme, which means that the enzyme is irreversibly inactivated. The 5-fluorouridine monophosphate is an example of a "**suicide substrate**"—a compound whose reaction with an enzyme causes the enzyme to no longer function.

Figure 6-5

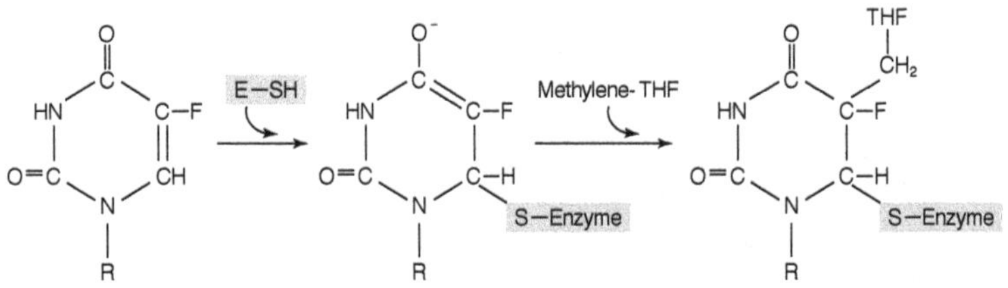

Another way to reduce the supply of deoxynucleotides for cell replication is to target the reduction of dihydrofolate to tetrahydrofolate. Folate antagonists are used in antimicrobial and anticancer chemotherapy. These compounds are competitive inhibitors of dihydrofolate reductase because they resemble the natural substrate. For example, methotrexate is used in antitumor therapy.

Figure 6-6

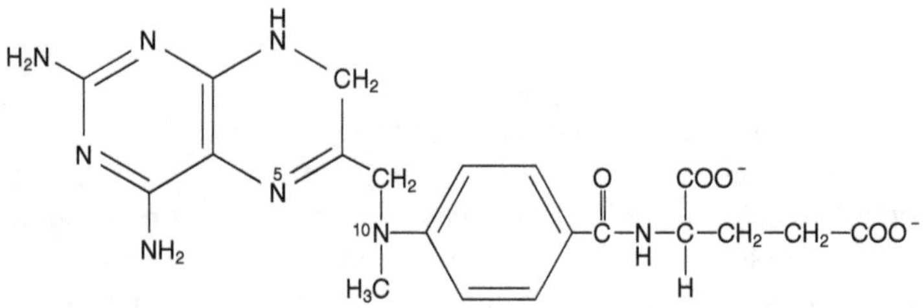

Folate antagonists can be overcome by increasing the amount of dihydrofolate reductase in the tumor cells. These resistant cells increase the amount of the enzyme by amplifying the DNA sequence encoding the enzyme. More copies of the gene make more molecules of enzyme. You can understand this by remembering that the Michaelis-Menten equation gives the velocity of an enzyme.

$$v = V_{max}[S]/K_M + [S]$$

$$V_{max} = k\ [E_t]$$

The velocity of a reaction is therefore proportional to the total amount of enzyme (E_t) in the previous reaction. If the activity of dihydrofolate reductase were to be inhibited tenfold by methotrexate, increasing E_t tenfold can restore the velocity (amount of tetrahydrofolate made per time).

CHAPTER 7
INTEGRATED METABOLISM

Metabolic Relationships

Metabolism can be overwhelmingly complex. A large number of reactions go on at the same time and remembering how they fit together with any kind of logic is difficult. The temptation to simply memorize the individual reactions is great, but some underlying principles can help your understanding. This chapter considers the ways in which mammalian (mostly human) metabolism fits together under different physiological conditions.

Energy from Glucose

Glucose is a preferred source of metabolic energy in almost all tissues. Glucose can be metabolized either aerobically or anaerobically. Although more energy is available by oxidative metabolism, some tissues can use glycolysis for a rapid burst of energy. The sources of glucose vary in different tissues.

Muscle is the largest consumer of glucose during exercise and can get glucose either from the circulation or by breaking down internal glycogen reserves. Two types of skeletal muscle exist and are distinguished by their physiological properties and their glucose metabolism.

In *fast white* fibers, glycolysis catabolizes glucose. The relative lack of mitochondria in these fibers causes the white appearance. The rapid breakdown of glucose by anaerobic metabolism means that ATP is made rapidly. These muscles are used in rapid, short-duration movement and exhibit a fast twitch when electrically stimulated. The flight muscles of birds are of this type—remember that you find the white meat of a chicken on the breast.

In *slow red* fibers, glucose metabolism leads into the TCA cycle and metabolism is aerobic. The red appearance of these muscles comes from the large number of mitochondria in them—the iron-containing cytochromes and myoglobin give the tissue its red appearance. The leg muscles (dark meat) of birds are of this type.

The same distinctions hold in humans. Sprinters and marathon runners have different proportions of muscle fibers, and therefore different metabolisms. Sprinters have relatively more fast white fibers, and can run very rapidly, but not for long distances. Marathon runners, on the other hand, have more slow red fibers and can carry out aerobic metabolism for very long periods of time, although they can't go as fast. Well-trained, world-class runners may have as much as 90 percent of their leg muscle of one type or the other, depending on their sport. Some sports, such as basketball and soccer, involve both aerobic endurance and anaerobic sprinting; these athletes tend to have both types of muscle fiber. Untrained individuals have about 50 percent of each type. The relative contributions of training and heredity to each type of metabolism remain unknown, although both play some part.

The *brain* relies on the circulation for nutrients and is a chief consumer of glucose. The brain uses about 15 percent of the energy required for minimal maintenance of body functions (called the basal metabolic rate). Brain tissue doesn't store energy. Instead, the brain must rely on the circulation for its fuel supply. Not all molecules can be transported across the blood-brain barrier to be used for energy. One molecule that can cross the blood-brain barrier is glucose, the preferred fuel source for the brain. Brain tissue can also adapt to ketone bodies such as acetoacetate as a source of fuel.

The *liver* is the store for and dispenser of carbohydrates to the circulation. Glycogen phosphorylase/glycogen synthetase enzyme activities control the glycogen breakdown. Hormones such as epinephrine and glucagon lead to the breakdown of glycogen to glucose-1-phosphate. Phosphoglucomutase then converts glucose-1-phosphate to glucose-6-phosphate, which is then dephosphorylated to glucose as it travels to the bloodstream.

Patients who have a hereditary deficiency of glucose-6-phosphate phosphatase accumulate large granules of glycogen in their livers. The only fate for glucose-6-phosphate in these patients is conversion into glucose-1-phosphate and then into glycogen. All the glucose that is not directly metabolized flows "one way," into glycogen.

Proteins and Fatty Acids

Other sources of metabolic energy include proteins and fatty acids. Muscle can use fatty acids from adipose (fat-containing) tissues. The first reaction, in the fat globule, is the hydrolysis of triacylglycerols by lipase, to give free fatty acids and glycerol. Fatty acids move through the circulation when bound to **serum albumin**. Serum albumin is a general carrier of molecules in the blood, but it seems to have a specific affinity for fatty acid. After it enters the muscle, β-oxidation breaks down the fatty acid to Acetyl-CoA molecules, which are then metabolized through the TCA cycle and the mitochondrial electron chain. Muscle can also use amino acids derived from proteolysis as energy sources, after their conversion to TCA cycle intermediates or pyruvate. The amino groups must be eliminated, preferably by converting them into urea in the liver. The "glucose-alanine" cycle transports amino groups from protein breakdown to the liver. Glucose from the liver circulates to muscle and is degraded to pyruvate; pyruvate is transaminated to alanine; alanine circulates back to the liver. In the liver, the amino group is converted to urea, leaving pyruvate, which can be reconverted to glucose. No net carbon metabolism occurs during this cycle—it moves amino groups only.

The *heart* oxidatively metabolizes a variety of substrates, probably because it is the most essential organ in the body. The energy demands of the heart are such that it probably must rely on ATP generated in the mitochondria; not enough ATP can be made anaerobically to support these demands. The heart efficiently metabolizes fatty acids and ketone bodies and may prefer those sources of energy, even to glucose.

Exercise and Metabolism

Exercise and prolonged fasting alter metabolic activity. Sprinting demands a quick input of energy. The first energy source for sprinting is the compound creatine phosphate. Like arginine, creatine contains a guanido group. Creatine is a muscle storage system for high-energy phosphate bonds. The ΔG^0 for the hydrolysis of creatine phosphate is about 10.3 kcal (43.1 kJ) per mole, which is about 3 kcal per mole greater than the hydrolysis of ATP. The reaction looks like this:

$$Cr + ATP \rightleftharpoons ADP + Cr{\sim}P$$

In the standard state, the equilibrium for this reaction lies far to the left; in other words, the reaction is *unfavored*. However, in the standard state, all the reactants and products are at one molar concentration. In other words, the ratio of ATP to ADP concentrations would be 1. In an actively metabolizing state, the ratio of ATP to ADP is as much as 50 or 100 to 1—this means that the formation of Cr~P will occur to a reasonable level. Creatine phosphate forms a reservoir for high-energy phosphate in the same way that water can be pumped upstream to a reservoir and released for use later on.

Anaerobic exercise: sprinting

During sprinting, the following series of events occur:

1. ATP is depleted as the muscles contract. ADP concentrations rise.
2. Phosphates are transferred back to ATP from creatine phosphate for further rounds of muscle contraction.

3. Creatine phosphate stores several times the amount of energy that is in ATP.
4. Quick ATP synthesis supplies the energy for a few seconds of sprinting.
5. Anaerobic glycolysis must work next.
6. Glucose comes from glycogen stores in muscle; it is catabolized to lactate and released into the circulation. As the ATP decreases in the muscle, the enzyme myokinase interconverts the resulting ADP to salvage one ATP out of two ADP.

$$\text{ADP} + \text{ADP} \rightleftharpoons \text{ATP} + \text{AMP}$$

The ATP can then power another contraction. Eventually, the amount of ATP available approaches a level too low to be bound by myosin in the muscle, even though it is by no means exhausted. The protons (acid) from metabolism cause hemoglobin to release its oxygen more readily, promoting a switch to aerobic metabolism. Lactate and protons from glycolysis may also lead to fatigue and an inability to sustain the level of speed that was possible earlier. In most humans, this seems to occur after a run of about 400 meters, which is why "running quarters" is one of the most unpleasant exercises for any athlete, no matter how well conditioned.

Aerobic metabolism: prolonged exercise

Only a finite amount of glycogen remains available in humans for exercise. The total glycogen plus glucose is about 600 to 700 kcal, even if it is metabolized aerobically. Running consumes this amount of energy in 1.5 to 2 hours. Using fat and/or muscle protein is necessary to keep going. Distance running therefore requires mobilization of fat to supplement glycogen breakdown; this is better done sooner than later in a long run. Competitive runners are dedicated to a variety of nutritional strategies alleged to mobilize fatty reserves.

Caffeine inhibits the breakdown of cyclic AMP and therefore contributes to the activation of glycogen phosphorylase; it also acts like epinephrine or glucagon and mobilizes lipids. Carnitine, the carrier of fatty acids into the mitochondrial matrix, is used as a dietary supplement with some success. Finally, distance runners try to increase their stores of muscle glycogen by **carbohydrate loading**. This is a two-step procedure. In the first step, a distance runner eats a very low-carbohydrate diet and exercises vigorously to deplete glycogen stores. Then he or she eats a large amount of carbohydrates, such as bread and pasta, which cause the muscles to store glycogen in greater amount than they would normally store.

Nutritional state

Humans usually eat a few times a day. This means that an individual's normal nutritional status cycles between two states, well-fed and fasting. Biochemically, the source of glucose, which is far and away the preferred source of energy for the brain, defines these states. During the well-fed state, the diet supplies glucose and the rest of the energy needed for protein synthesis. Between meals, the breakdown of glycogen and gluconeogenesis from amino acids supplies the glucose requirements. In more advanced cases of starvation, muscle protein is broken down more extensively for gluconeogenesis. In advanced stages of starvation, glucose metabolism is reduced and the brain metabolizes ketone bodies for energy.

The digestion of foodstuffs in and absorption from the intestine characterizes the **well-fed state**, which lasts for about four hours after a meal. Free amino acids and glucose are absorbed and transported to the liver. Excess energy is converted to fat in the liver and transported, along with dietary fat, to the adipose tissues. The pancreas releases high levels of *insulin* in response to these events. Insulin signals the liver to convert glucose to glycogen, amino acids to protein, and fat to triglycerides. The adipose tissues synthesize and deposit fats. A deficiency of insulin is a cause of *diabetes*, characterized by excess levels of blood glucose. In this disease, glucose is not converted into glycogen or fat, so it remains in the circulation.

As the individual enters the **fasting state**, glycogen is broken down into glucose to supply energy for the tissues. Simultaneously, gluconeogenesis begins as amino acids, lactate, and pyruvate from metabolism are cycled into the formation of glucose. As fats are broken down, the fatty acids supply energy to the peripheral tissues, while the glycerol from breaking down the triacylglycerols is transported to the liver and converted to glucose. Gluconeogenesis becomes more important than glycogen breakdown after about 16 hours of fasting. Gluconeogenesis is maximal after about two days without food, at which time ketone bodies are made from fat and transported to the brain. This transition describes the beginning of *starvation*, which can last for six to ten weeks before death occurs. During starvation, the body breaks down amino acids for glucose; however, ketone bodies and fat supply most energy requirements. At this time, the body is in *negative nitrogen balance*, because the amount of nitrogen excreted due to protein breakdown exceeds the nitrogen eaten in food. The small amount of glucose made is supplied to brain, kidney, and red blood cells. The latter two tissues have no alternative energy sources; the brain uses both ketone bodies and glucose. When fat is gone, the only sources of energy available are amino acids from muscle. The carbon skeletons are metabolized, and the nitrogen is excreted. This situation cannot continue for very long. Eventually, the kidneys fail, or the heart muscle is broken down, and the individual dies.

Hormonal Regulation

Hormones regulate metabolic activity in various tissues. They are one kind of mechanism for signaling among cells and tissues. Hormones can be defined as signaling molecules that one cell releases into the peripheral fluid or bloodstream, which alter the metabolism of the same or another cell. Hormones are distinguished from communication mechanisms that depend on direct cell-cell contact through gap junctions. Hormones are also distinguished from neurotransmitters, although this distinction is somewhat artificial. Neurotransmitters can act as hormones and vice versa.

Hormones act by binding to **receptors**, which are usually protein molecules. Receptors have two functions: first, they *bind* the hormone, and secondly, they *transduce* (change the type of) the signal to affect the metabolism of the recipient cell. The ability of a cell to respond to a hormone depends on two properties of the receptor molecule: how many of them are on a particular cell, and how well they bind the hormone. The first property is called the **receptor number**, and the second is called the **affinity** of the receptor for the hormone. The biochemical responsiveness of a cell to a hormone (or a drug, or a neurotransmitter) depends on the *number of occupied receptors* on the responsive cell. Suppose that a hormone binds to a receptor with a dissociation constant given by the following equation:

$$K_d = [R][H]/[RH]$$

In the equation, R is the receptor, H is the hormone, and RH is the hormone-receptor complex. If 50 occupied receptors trigger the appropriate metabolic response, you can achieve the response by having 100 receptors on a cell with half of them occupied or by having 55 receptors on a cell with 90 percent of them occupied. How can this be achieved? If the second set of receptors had a tenfold greater affinity for the hormone, the same concentration of hormone would result in 50 bound receptors.

Rearranging the previous equation to solve for [H], the level of circulating hormone yields

$$[H] = K_d\ [RH]/[R]$$

If two receptors exist, types 1 and 2, each of which is responding to a constant concentration of hormone, [H], then

$$[H] = K_1\ [RH_1]/[R_1] = K_2\ [RH_2]/[R_2]$$

$$10\ K_2 = K_1$$

(Remember that the higher K_d means a lower affinity.)

If you set the number of occupied receptors $[RH_1] = [RH_2] = 50$, you can solve for the number of unoccupied receptors of each type $[R_1] = 50$, and $[R_2] = 5$. In other words, one receptor type has a greater *occupancy* than the other does.

Suppose the hormone concentration increased by 50 percent. In this case, the first receptor system, R_1, would be more responsive. R_2 would be close to **saturation**; complete saturation of R_2 would yield only five more occupied receptors. This means that the concentration of occupied receptors can change most when the receptor is about half occupied. The previous equations show that *the maximum responsiveness to a change in hormone concentration is possible when the association constant of the receptor for a hormone is near the physiological concentration of the hormone.*

The compounds that bind to a receptor can modulate the actions of that receptor. **Agonists** act to reinforce the activity of a receptor by binding to it and mimicking the action of the receptor. **Antagonists** bind to a receptor but do not cause the action of the receptor. Drugs can be either agonists or antagonists. For example, isoproteranol is an agonist for a receptor that increases blood pressure, while propanolol—a commonly used drug to decrease blood pressure—is an antagonist for another class of receptors. Both of these compounds are structurally related to the natural hormone epinephrine.

Down regulation

When cells are continually occupied, they reduce the number of receptors to avoid having the metabolic effects overstimulated. For example, two kinds of diabetes exist, Type I and Type II. Type I diabetes, sometimes called juvenile diabetes, results from the inability of the pancreas to supply insulin. Type II diabetes, sometimes called adult-onset diabetes, is more common and correlates with obesity. In this situation, the body senses itself to be in a well-fed state and releases insulin from the pancreas. The large concentration of insulin causes the recipient cells to be fully stimulated. Consequently, they down-regulate their insulin receptor population to bring the response into the normal range. Reducing the total number of receptors reduces the number of occupied receptors. Unfortunately, the lower number of occupied receptors means that blood glucose concentrations are not well regulated and can increase to an abnormally high level, leading to the side effects of diabetes, including problems with vision and circulation.

Receptor Types

Hormone receptors are of two types. **Cell-surface receptors** for water-soluble hormones lead to a metabolic response. For example, receptors for the small molecule epinephrine and for the peptide growth hormone are of this type. A protein in the cell membrane binds the hormone and then causes the synthesis of a **second messenger**, which leads to a metabolic response.

Internal or **cytoplasmic receptors** for lipophilic hormones lead to gene activation. For example, the steroid hormones, including estrogen and adrenal hormones, bind to intracellular receptors. When occupied, these receptors then move to the cell nucleus and affect transcription of specific genes, either positively or negatively. Some hormones—including insulin—exert both metabolic and genetic changes when bound to surface receptors.

Metabolic hormones do not exert their effects directly but rather are transduced into an intracellular signal. **Transduction** refers to the process by which one kind of signal is converted to another.

Cyclic AMP: A Second Messenger

The action of epinephrine illustrates the principles by which cyclic AMP mediates hormone action. Epinephrine is the "flight or fight hormone" that the adrenal glands release in response to stress. The hormone causes an increase in blood pressure and the breakdown of glucose for energy. This helps humans in danger to engage in physical activity to meet the challenges of a situation. The body responds with a dry mouth, rapid heartbeat, and high blood pressure. A biochemical chain of events leads to these responses.

When epinephrine binds to cells, it stays outside on the membrane-bound receptor. The second messenger, cyclic AMP, is made by the enzyme **adenylate cyclase**.

Cyclic AMP:

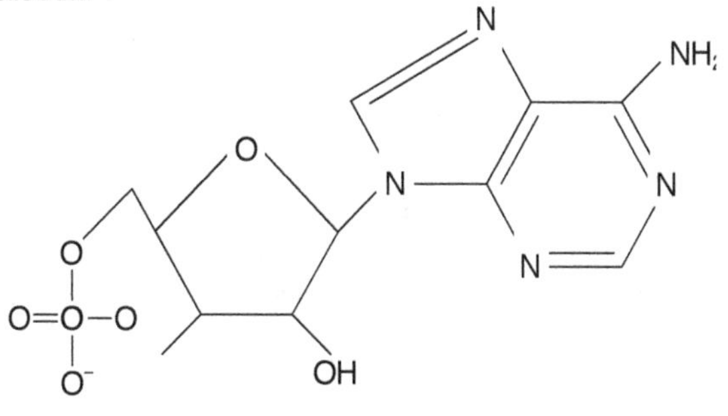

Adenylate cyclase is a two-component enzyme system. It ultimately catalyzes the cyclase reaction, but only when it is associated with the hormone-bound receptor and a regulatory protein called a stimulatory **G-protein** (guanylate nucleotide binding protein), which activates adenylate cyclase. The G-protein is the intermediate between the receptor and the synthesis of cyclic AMP.

G-proteins exist either in an active or an inactive state, depending on the guanylate nucleotide that is bound. In the inactive state, G-protein binds to GDP. In the active state, GTP is bound to the G-protein. G-proteins have an intrinsic **GTPase** activity, which converts bound GTP to GDP. Hydrolysis of GTP by the G-protein converts the G-protein back to an inactive state. Thus the cycle of the G-protein is as follows:

1. Hormone binds to receptor.
2. The hormone-bound receptor binds to the G-protein and causes GDP to be replaced by GTP.
3. GTP-bound G-protein interacts with adenylate cyclase.
4. G-protein hydrolyzes bound GTP to GDP, thereby going back to the ground state.

Figure 7-1

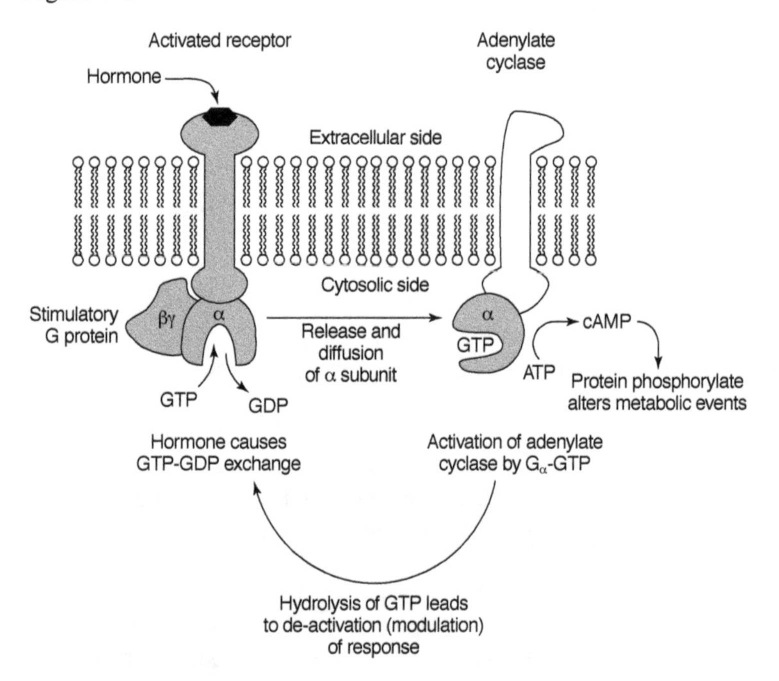

Different G-proteins may either stimulate or inhibit adenylate cyclase to make more or less cyclic AMP.

Action of cyclic AMP

Cyclic AMP doesn't act directly on its target enzymes; for example, glycogen phosphorylase and glycogen synthase. Instead, cyclic AMP stimulates a *protein kinase* cascade that ultimately leads to a cellular response. Cyclic AMP binds to *protein kinase A*, which then catalyzes the transfer of phosphate from ATP to a serine residue on a second enzyme, *phosphorylase kinase*, which itself transfers a phosphate to glycogen phosphorylase. Active glycogen phosphorylase then catalyzes the breakdown of glycogen to glucose-1-phosphate. This provides energy for muscle activity.

Cells can't be "turned on" forever. Something must *modulate* the response. In fact, each step is reversible. Starting from the target proteins, a *protein phosphatase* hydrolyzes the phosphate from the proteins. Cyclic AMP is hydrolyzed by a *phosphodiesterase*.

Perhaps a key point in the modulation system is GTP hydrolysis by the G-protein. This causes adenylate cyclase to return to the unstimulated state.

All signaling mechanisms must have this modulation feature to allow the possibility of control. For example, the Ras protein of mammalian cells is a membrane-bound GTPase. Mutations that decrease Ras's GTPase activity can contribute to uncontrolled growth (i.e., tumor formation) of mammalian cells.

PI System: Another Second Messanger

The phosphatidylinositol (PI) system is another second messenger system. PI is a minor component of membrane lipids. This molecule serves as a source of second messenger compounds. PI has three parts. See Figure 7-2.

Figure 7-2

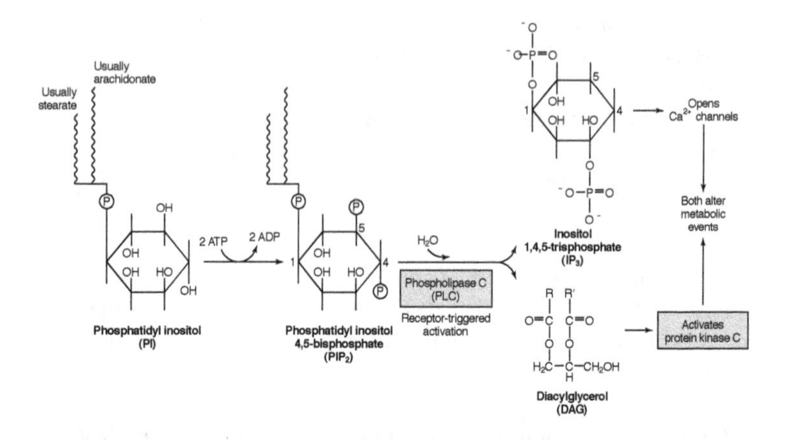

The first part of PI consists of two fatty acids esterified to a glycerol. One of the fatty acids is the unsaturated fatty acid *arachidonic acid* (20:4), bound to carbon 2 of the glycerol. The other fatty acid is usually *stearate* (18:0). The combination of two fatty acids esterified to glycerol is called **diacylglycerol**, abbreviated **DAG**.

Another component of PI is a carbohydrate, **phosphoinositol**, which a phosphate diester binds to the third position of the glycerol. The inositol is usually phosphorylated at two positions.

Several types of signaling molecules are derived from PI. Hydrolysis of the glycerol-phosphate linkage by a *phospholipase* leads to the signaling molecule **trisphosphoinositol**, abbreviated **IP_3**.

PO OH
OP
OH
PO
OH

The remaining part of the phosphoinositol, 1,2-diacylglycerol, is also a signaling molecule.

Finally, the arachidonate that arises from cleavage of phosphatidylinositol can serve as a precursor of *prostaglandins*.

Figure 7-3

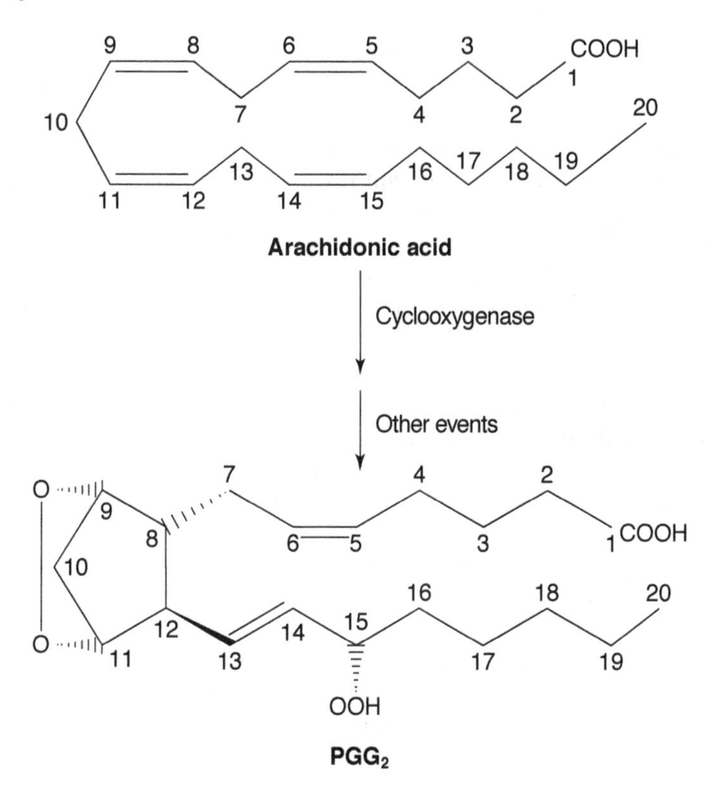

Prostaglandins are mediators of the inflammatory response and are produced by the action of two activities of the enzyme prostaglandin synthase. The first activity is a **cyclooxygenase** activity, which adds two oxygen molecules the arachidonic acid. Secondly, the peroxide group from the first step is reduced to a hydroxyl group.

The drug acetylsalicylic acid (aspirin) irreversibly inhibits the cyclooxygenase activity, while ibuprofen inhibits the reductase activity. Both drugs treat inflammation, pain, and fever because they inhibit prostaglandin synthesis. Prostaglandins are very unstable, so they tend to act locally (otherwise a sprained ankle would cause pain throughout the body).

IP_3 mobilizes Ca2+ from intra- or extracellular stores. The interior of a cell is kept very low in Ca2+ ions, at a concentration less than 10-9 M., while the outside [Ca2+] is about 10-3 M. This million-fold concentration gradient is the result of cellular calcium-dependent ATPase protein. Ca-ATPase uses up to a third of the ATP synthesized by a cell to maintain the concentration gradient. The stores of Ca2+ available for use inside the cell are found primarily in the endoplasmic reticulum. A large store of Ca2+ exists in the mitochondrial matrix, but this seems to be a final "dumping ground"—in other words, calcium ions in the mitochondria don't come into the cytoplasm.

After Ca2+ comes into the cytoplasm, it binds to the mediator protein **calmodulin**. Calmodulin is a subunit of phosphorylase b kinase and a number of other enzymes. It binds Ca2+ with a K_d of approximately 10^{-6} M. When it does, calmodulin undergoes a conformational change; this conformational change activates phosphorylase kinase, which in turn leads to the activation of glycogen breakdown. Thus, an increase in intracellular [Ca2+] acts in the same manner as cyclic AMP. Phosphatases in the cell rapidly hydrolyze IP_3, which modulates the signal. The cytoplasmic Ca2+ is transported to the mitochondria and the cell returns to the resting state. The inositol is reincorporated into lipid and then re-phosphorylated, ready to serve as a source of second messengers again.

Protein kinase C

After the phosphoinositol is released from the phosphatidylinositol, the remaining diacylglycerol is itself a second messenger, activating protein kinase C. The targets of protein kinase C are a number of intracellular proteins, including some that apparently target cell growth. Inappropriate stimulators of protein kinase C can be tumor promoters, compounds that stimulate the growth of cancer cells but do not cause genetic changes in the cell's DNA.

Receptors with Kinase Activity

Hormone binding can directly stimulate the protein kinase activity of receptors. The insulin receptor is an example of this type. The small protein insulin binds to its receptor, which crosses the outer membrane of a cell. These proteins have three domains. The **extracellular** domain of these receptors binds the hormone, the **transmembrane** domain crosses the membrane, and the **intracellular** domain is a protein kinase. The activity of the kinase domain is stimulated when the hormone is bound to the receptor. See Figure 7-4.

Another protein hormone, epidermal growth factor, is bound to the extracellular domain of its receptor. A mutated form of the receptor, which lacks the hormone-binding domain, results in the kinase activity being permanently "on." This leads to cancer because the cell is signaled to grow at all times—the hallmark of cancer.

Figure 7-4

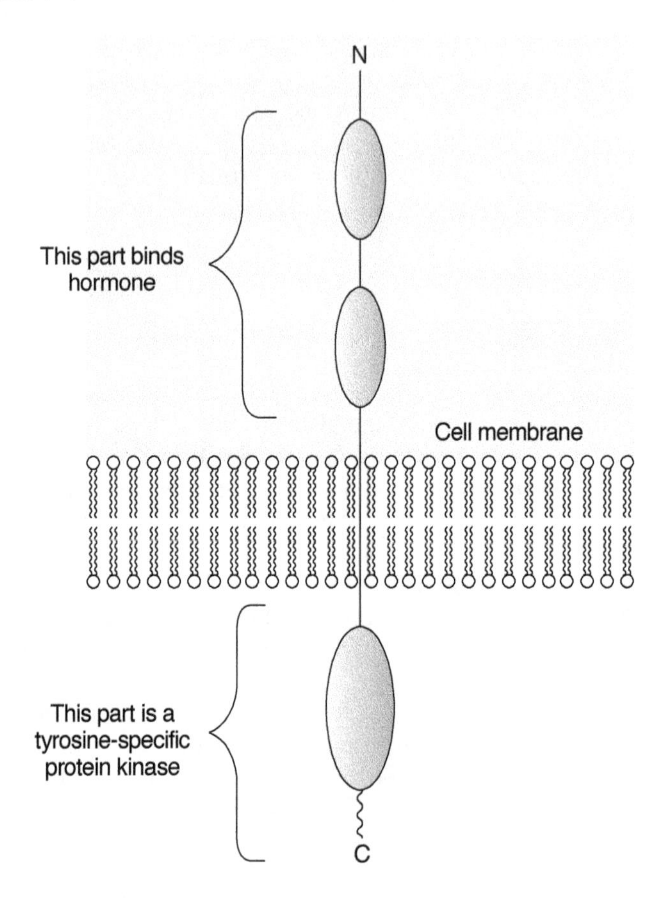

Hormone interactions
Responses to various hormones can "cross-talk" by means of multiple interactions/modifications. One example has already been discussed: Phosphorylase b kinase is responsive both to cAMP and to Ca^{2+} ions. Receptors can themselves be phosphorylated by protein

kinases, which can change their activity. Thus, a protein kinase may respond to cyclic AMP (for example, through the epinephrine receptor), intracellular calcium ion concentration (through the IP_3 system), and an extracellular hormone such as a growth factor (through the kinase activity of a receptor). All of these activities can reinforce or antagonize each other, providing precise control.

Hormones that Affect Gene Activity

Lipid-soluble hormones act usually by gene activation/deactivation. Examples of these hormones include steroids, thyroid hormone, and vitamin A (retinoic acid). The hormones are transported through the circulation in association with a hormone-binding protein and are soluble in the plasma membrane of the cell. Their receptors are *intracellular*, and they act on *gene transcription* (the synthesis of messenger RNA) rather than at the protein level. Thus, they act more slowly than do the soluble hormones, on the scale of days rather than minutes.

The sequence of events in gene activation contains several steps. First, in the hormone-free state, the unoccupied receptor is bound to the nuclear membrane and loosely to **chromatin**. (Chromatin is the DNA-protein complex of chromosomes.) After the hormone binds the receptor, it changes its location. The receptor-hormone complex binds DNA tightly and thereby activates or inactivates the synthesis of mRNA from these genes. The specificity of these receptors lies in two properties: their ability to bind different hormones and their ability to bind different DNA sequences.

These receptors share quite similar DNA-binding domains and differ somewhat more in their hormone-binding regions as shown in Figure 7-5. Additionally, they have very different **activation domains**, which interact with other parts of the transcription machinery.

Figure 7-5

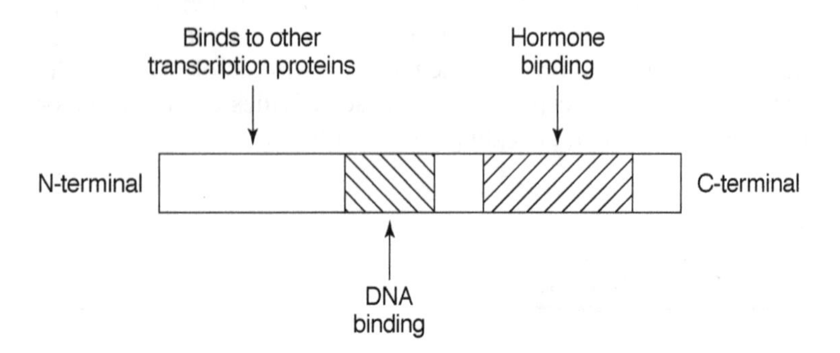

Again, the possibility of "cross-talk" exists between metabolic and genetic events. Thus, for example, steroids may bind to one receptor, which itself will interact with other proteins. Some of these proteins may be phosphorylated by kinases that respond to the presence of cAMP or a Ca2+ ion.

CHAPTER 8
DNA STRUCTURE, REPLICATION, AND REPAIR

DNA and RNA Structures

Nucleic acids have a primary, secondary, and tertiary structure analogous to the classification of protein structure. The sequence of bases in the nucleic acid chain gives the primary structure of DNA or RNA. The sequence of bases is read in a 5' → 3' direction, so that you would read the structure in the next figure as ACGT. See Figure 8-1.

The base-pairing of complementary nucleotides gives the secondary structure of a nucleic acid. In a double-stranded DNA or RNA, this refers to the Watson-Crick pairing of complementary strands. In a single-stranded RNA or DNA, the *intramolecular* base pairs between complementary base pairs determines the secondary structure of the molecule. For example, the cloverleaf structure of Figure 8-2a gives the secondary structure of transfer RNAs.

The regions of the secondary structure do not have to form between sequences that are close together. For example, base-pairing between the sequences at the 5' and 3' ends forms the **acceptor stem** of transfer RNA. The tertiary structure of a nucleic acid refers to the three-dimensional arrangement of the nucleic acid—that is, the arrangement of the molecule in space, as in the tertiary structure of tRNA.

Figure 8-1

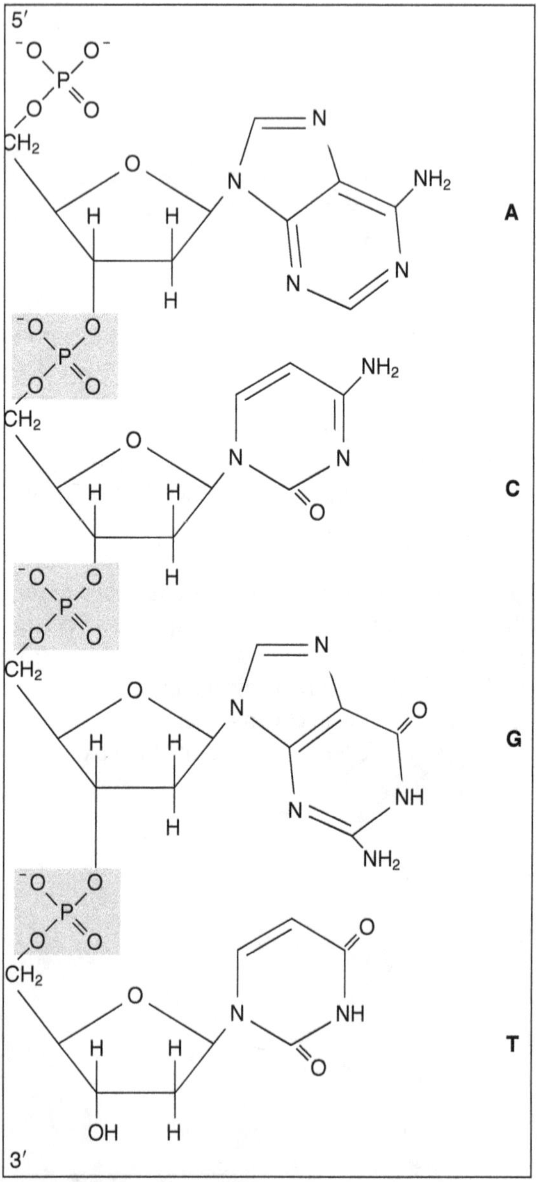

5′
^-O
O^-
P
O
O
CH_2
O
N
N
NH_2
H
H
N
N
H
A
^-O
O
P
O
O
NH_2
CH_2
O
N
N
C
H
H
O
H
^-O
O
P
O
O
N
CH_2
O
N
O
G
H
H
NH
N
H
NH_2
^-O
O
P
O
O
O
CH_2
O
N
NH
T
H
H
O
OH
H
3′

Figure 8-2a

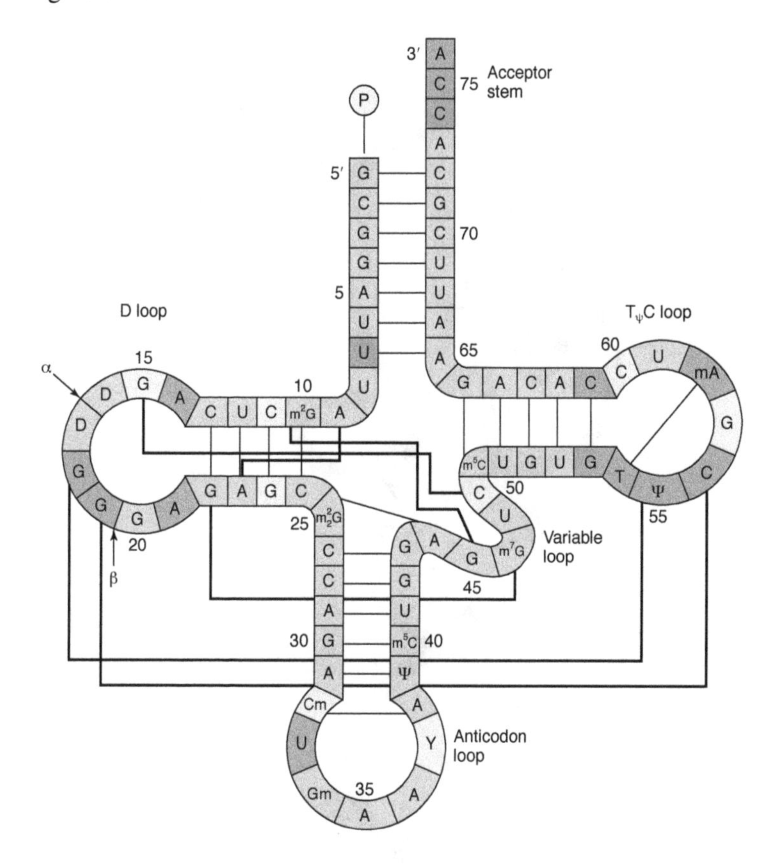

Secondary-Base pair

Figure 8-2b

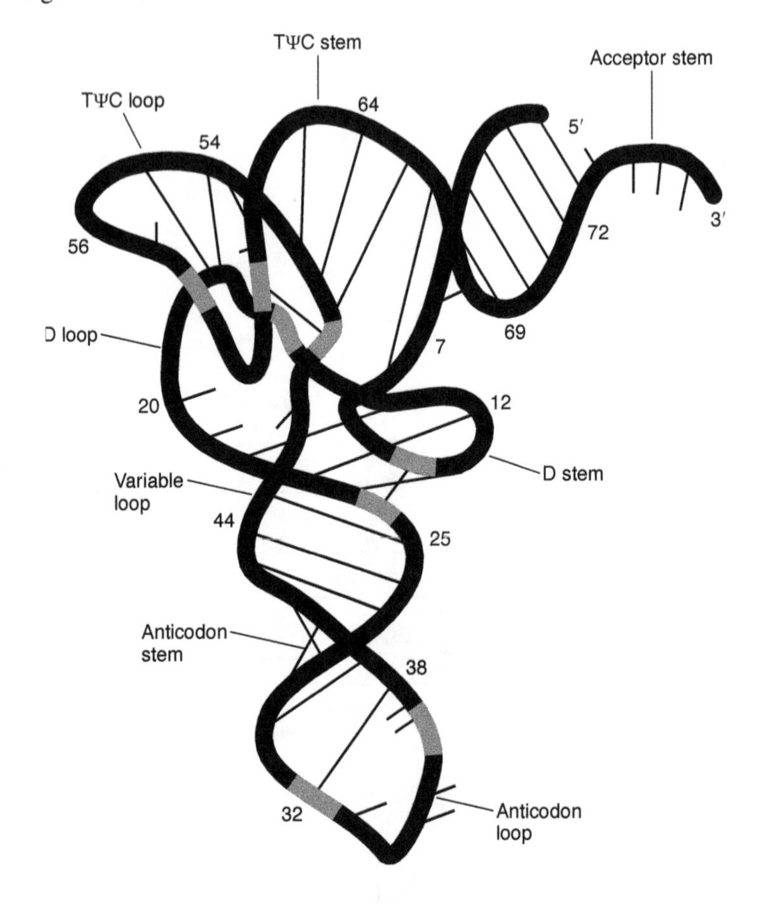

Tertiary = Folded in 3 dimensions

Sugar structures in DNA and RNA

Although RNA and DNA are similar in their overall properties, the presence of the 2'-hydroxyl group of ribose in RNA and its replacement by 2'-deoxyribose in DNA make a large contribution to the different chemical properties of DNA and RNA. The 2'-hydroxyl group makes RNA susceptible to *alkaline hydrolysis* in solutions of pH greater than nine or so. Conversely, the bonds between the sugar and the purine bases of DNA are more easily broken in low pH conditions than are those of RNA.

The 2'-hydroxyl group affects the tertiary structure of RNA. First, the conformation of the sugar is different between DNA and RNA. Secondly, the 2'-hydroxyl group provides hydrogen bond donor and acceptor functions for formation of hydrogen bonds. These hydrogen bonds are important in the formation of the tertiary structure of an RNA and are not available to DNA. Although single-stranded DNA does have some tertiary structure, this structure is usually not as stable as that of an RNA of the same sequence.

DNA structure

The Watson-Crick base-pairing of the two strands largely determines the **secondary structure of DNA**. All naturally occurring DNAs are double-stranded, for at least some of their lifetimes. Double-stranded DNA is a fairly uniform structure, and the need for a regular structure is one way in which changes in DNA (genetic mutations) can be detected. The fact that A-T base pairs and G-C base pairs have very similar sizes means that no "bulges" or "gaps" exist within the double helix. An irregular place in the double helix means that something is wrong with the structure, and this signals the need for **DNA repair systems** to fix the damage.

The A-T base pair has two hydrogen bonds; each base serves as H-donor for one bond and as H-acceptor for the other.

The G-C base pair has three hydrogen bonds; G is an acceptor for one for these, and a donor for two. This has important consequences for the **thermal melting** of DNAs, which depends on their base composition.

Figure 8-3

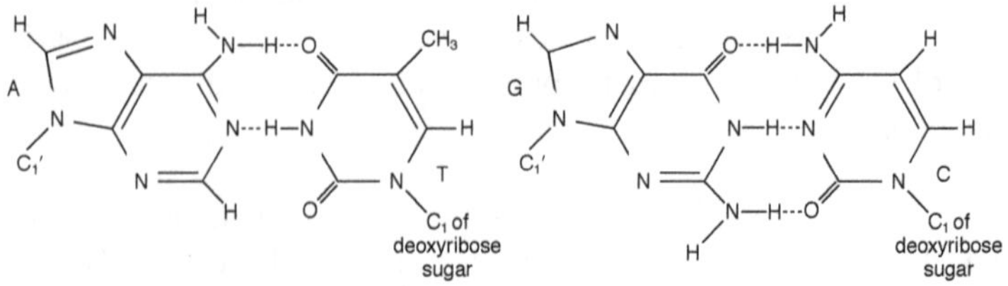

Thermal melting refers to heating a DNA solution until the two strands of DNA separate, as shown in Figure 8-4. Conversely, a double-stranded molecule can be formed from complementary single stands.

Melting and helix formation of nucleic acids are often detected by the **absorbance of ultraviolet light**. This process can be understood in the following way: The stacked bases shield each other from light. As a result, the absorbance of UV light whose wavelength is 260 nanometers (the A_{260}) of a double-helical DNA is less than that of thc same DNA, whose strands are separated (the random coil). This effect is called the **hypochromicity** (less-color) of the double-helical DNA.

If a double-stranded DNA is heated, the strands separate. The temperature at which the DNA is halfway between the double-stranded and the random structure is called the **melting temperature (T_m)** of that DNA. The T_m of a DNA depends on base composition. G-C base pairs are stronger than A-T base pairs; therefore, DNAs with a high G+C content have a higher T_m than do DNAs with a higher A+T content. For example, human DNA, which is close to 50 percent G+C, might melt at 70°, while DNA from the bacterium *Streptomyces,* which has close to 73 percent G+C, might melt at 85°. The T_m of a DNA also depends on solvent composition. High ionic strength—for example, a high concentration of NaCl—promotes the double-stranded state (raises the T_m) of a given DNA because the higher concentration of positive sodium ions masks the negative charge of the phosphates in the DNA backbone.

Finally, the T_m of a DNA depends on how well its bases match up. A synthetic DNA double strand made with some mismatched base-pairs has a lower T_m compared to a completely double-stranded DNA. This last property is important in using DNA from one species to detect similar DNA sequences of another species. For example, the DNA coding for an enzyme from human cells can form double helices with mouse DNA sequences coding for the same enzyme; however, the mouse-mouse and human-human double strands will both melt at a higher temperature than will the human-mouse hybrid DNA double helices.

Figure 8-4

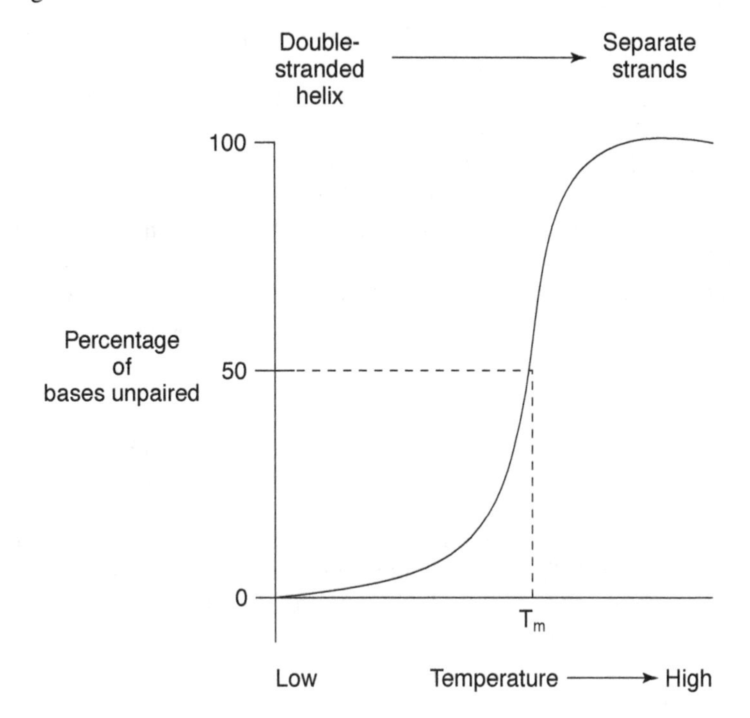

Direct reactions with DNA serve as the molecular basis for the action of several anti-tumor drugs. Cancer is primarily a disease of uncontrolled cell growth, and cell growth depends on DNA synthesis. Cancer cells are often more sensitive than normal cells to compounds that damage DNA. For example, the anti-tumor drug cisplatin reacts with guanine bases in DNA and the daunomycin antibiotics act by inserting into the DNA chain between base pairs. In either case, these biochemical events can lead to the death of a tumor cell.

DNA tertiary structure

The DNA double helix may be arranged in space, in a tertiary arrangement of the strands. The two strands of DNA wind around each other. In a **covalently closed circular DNA**, this means that the two strands can't be separated. Because the DNA strands can't be separated, the total number of turns in a given molecule of closed circular DNA is a constant, called the **Linking Number**, or **Lk**. The linking number of a DNA is an integer and has two components, the **Twist** (**Tw**), or number of helical turns of the DNA, and the **Writhe** (**Wr**), or the number of **supercoiled turns** in the DNA. Because L is a constant, the relationship can be shown by the equation:

$$Lk = Tw + Wr$$

Figure 8-5a and b, which shows a double helical DNA with a linking number equal to 23, best illustrates this equation.

Normally, this DNA would have a linking number equal to 25, so it is **underwound**. The DNA double helical structures in the previous figure have the same value of Lk; however, the DNA can be supercoiled, with the two "underwindings" taken up by the negative supercoils. This is equivalent to two "turns'-worth" of single-stranded DNA and no supercoils. This interconversion of helical and superhelical turns is important in gene transcription and regulation.

Figure 8-5a

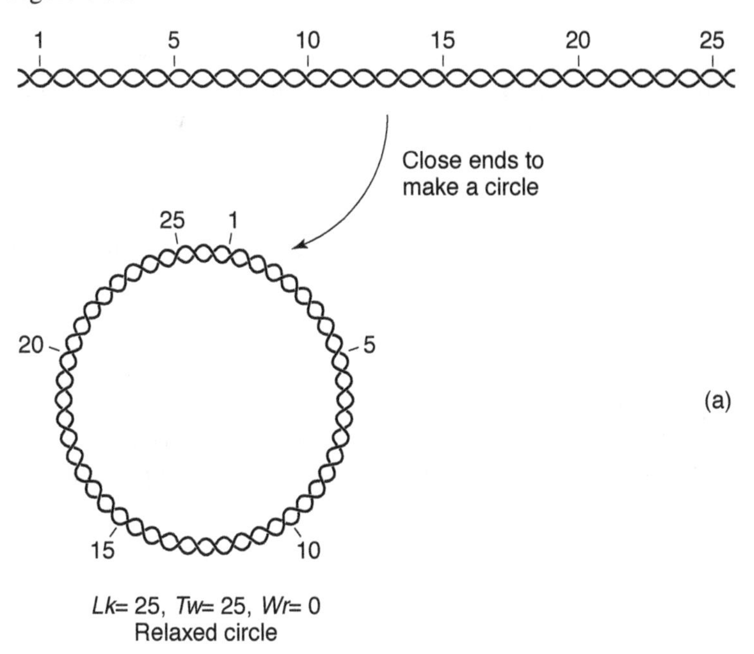

Figure 8-5b

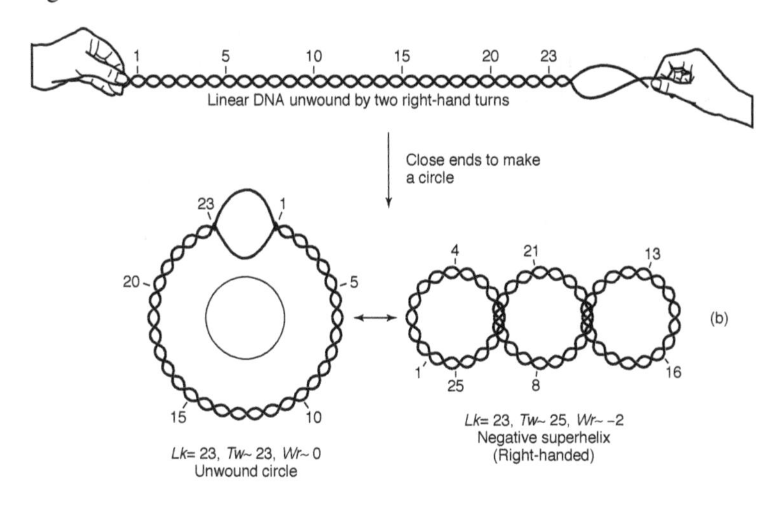

Enzymes called **DNA topoisomerases** alter Lk, the linking number of a DNA, by a bond breaking and rejoining process. Naturally-occurring DNAs have negative supercoils; that is, they are "underwound." **Type I** topoisomerases (sometimes called "nicking-closing enzymes") carry out the conversion of negatively supercoiled DNA to relaxed DNA in increments of one turn. That is, they increase Lk by increments of one to a final value of zero. Type I topoisomerases are energy independent, because they don't require ATP for their reactions. Some anti-tumor drugs, including campothecin, target the eukaryotic topoisomerase I enzyme. **Type II** topoisomerases (sometimes called DNA gyrases) reduce Lk by increments of two. These enzymes are ATP-dependent and will alter the linking number of any closed circular DNA. The antibiotic naladixic acid, which is used to treat urinary tract infections, targets the prokaryotic enzyme. Type II topoisomerases act on naturally occurring DNAs to make them supercoiled. Topoisomerases play an essential role in DNA replication and transcription.

DNA Replication Enzymes

Watson and Crick (see Volume 1) immediately saw the relationship of the double helix to genetic replication. They proposed that each strand of the chromosome serves as a **template** to specify a new, complementary DNA strand. A template is a pattern for making something; DNA acts as a template because each strand specifies the new **daughter** strand by base-pairing. This template feature makes DNA replication **semiconservative**: after replication, each **daughter chromosome** has one strand of newly synthesized DNA and one strand of DNA from the parental chromosome. See Figure 8-6.

Figure 8-6

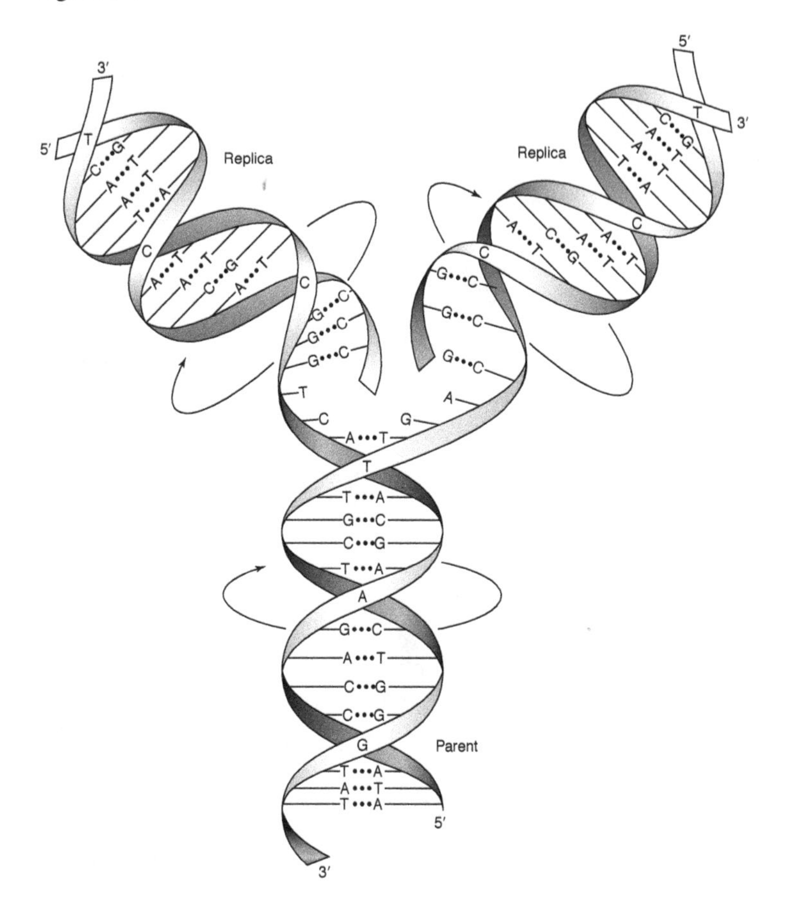

DNA polymerases

These enzymes copy DNA sequences by using one strand as a template. The reaction catalyzed by DNA polymerases is the addition of deoxyribonucleotides to a DNA chain by using dNTPs as substrates, as shown in Figure 8-7.

Figure 8-7

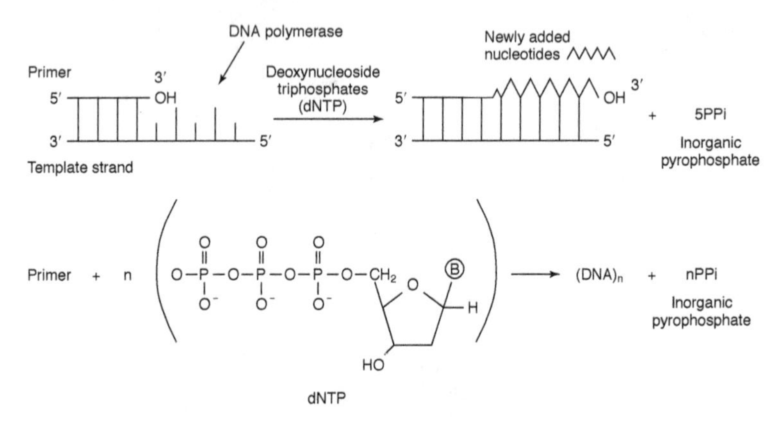

All DNA polymerases require a template strand, which is copied. DNA polymerases also require a **primer**, which is complementary to the template. The reaction of DNA polymerases is thus better understood as the addition of nucleotides to a primer to make a sequence complementary to a template. The requirement for template and primer are exactly what would be expected of a **replication** enzyme. Because DNA is the information store of the cell, any ability of DNA polymerases to make DNA sequences from nothing would lead to the degradation of the cell's information copy.

More than one DNA polymerase exists in each cell. The key distinction among the enzyme forms is their **processitivity**—how long a chain they synthesize before falling off the template. A DNA polymerase used in replication is more processive than a repair enzyme.

The replication enzyme needs to make a long enough chain to replicate the entire chromosome. The repair enzyme needs only to make a long enough strand to replace the damaged sequences in the chromosome. The best-studied bacterium, **E. coli**, has three DNA polymerase types.

DNA polymerase I (Pol I) is primarily a repair enzyme, although it also has a function in replication. About 400 Pol I molecules exist in a single bacterium. DNA polymerase I only makes an average of 20 phosphodiester bonds before dissociating from the template. These properties make good sense for an enzyme that is going to replace damaged DNA. Damage occurs at separate locations so the large number of Pol I molecules means that a repair enzyme is always close at hand.

DNA polymerase I has nucleolytic (depolymerizing) activities, which are an intimate part of their function. The 5' to 3' exonuclease activity removes base-paired sequences ahead of the polymerizing activity. During replication, this can remove primers ahead of the polymerizing function of the polymerase. See Figure 8-8.

Figure 8-8

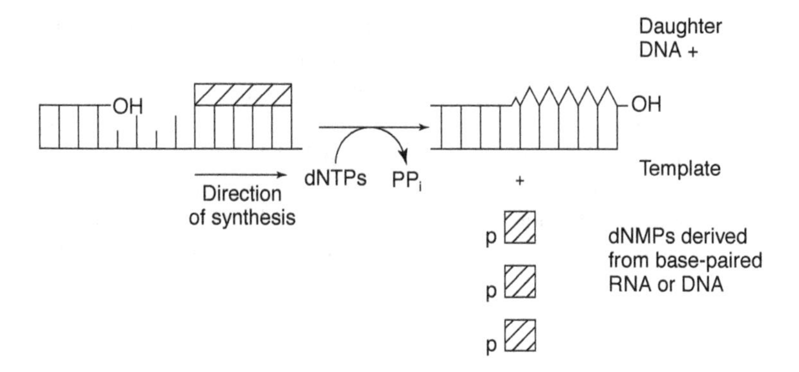

Another intimate function of DNA polymerase I (and of the other forms of DNA polymerase found in E. coli) is the 3' to 5' exonuclease activity. This activity can de-polymerize DNA starting from the newly synthesized end. Imagining why DNA polymerase would have an activity that opposes the action of the enzyme is a little difficult. The 3' to 5' exonuclease activity serves an **editing** function to ensure the fidelity of replication. Suppose DNA polymerase were to make a mistake and add a T opposite a G in the template strand. When the enzyme begins the next step of polymerization, the T is not properly paired with the template. The 3' to 5' exonucleolytic activity of DNA polymerase then removes the unpaired nucleotide, releasing TMP, until a properly paired stretch is detected. Then polymerization can resume. This cycle costs two high-energy phosphate bonds because TTP is converted to TMP. While this may seem wasteful of energy, the editing process does keep the information store of the cell intact, as shown in Figure 8-9.

DNA polymerase II is a specialized repair enzyme. Like Pol I, a large number of Pol II molecules reside in the cell (about 100). The enzyme is more processive than Pol I. Pol II has the same editing (3' to 5') activity as Pol I, but not the 5' to 3' exonuclease activity.

The actual replication enzyme in E. coli is **DNA polymerase III**. Its properties contrast with Pol I and Pol II in several respects. Pol III is much more processive than the other enzymes, making about 500,000 phosphodiester bonds on the average. In other words, it is about 5,000 times more processive than Pol I and 50 times more processive than Pol II. Pol III is a multisubunit enzyme. It lacks a 5' to 3' exonucleolytic activity, although a subunit of the enzyme carries out the editing (3' to 5') function during replication. Finally, only about 10 molecules of Pol III reside in each cell. This remains consistent with the function of Pol III in replication, because the chromosome only needs to be copied once per generation. Therefore, the cell only requires a few molecules of the enzyme. Pol III synthesizes DNA at least a hundred times more rapidly than the other polymerases. It can synthesize half of the bacterial chromosome in a little more than 20 minutes, which is the fastest that the bacterium can replicate.

Figure 8-9

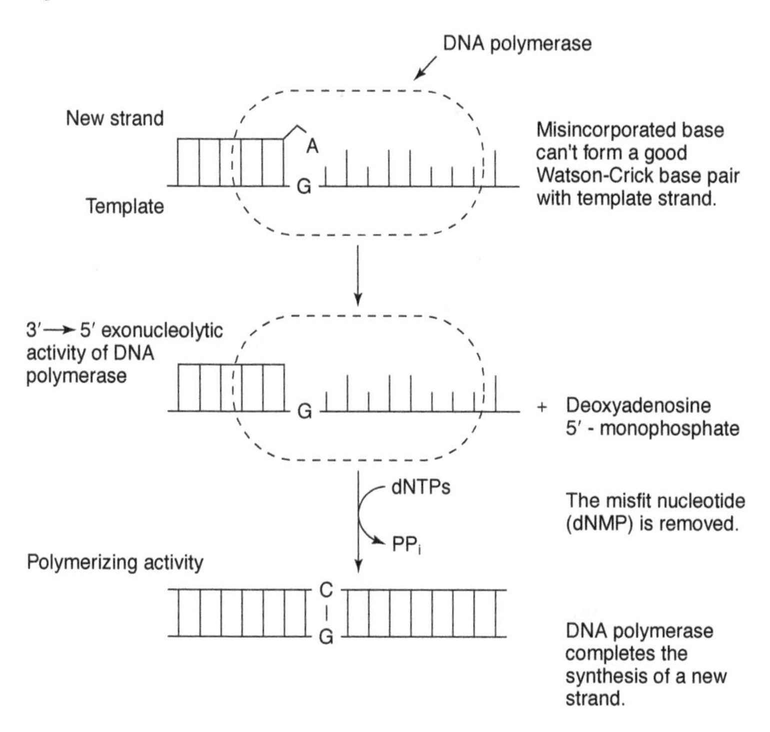

Chromosomal replication

The process of chromosomal replication in bacteria is complex. Bacterial chromosomes are double-stranded DNA and almost always circular. DNA replication starts at a specific sequence, the **origin**, on the chromosome and proceeds in two directions towards another specific region, the **terminus**, as shown in Figure 8-10a.

Figure 8-10a

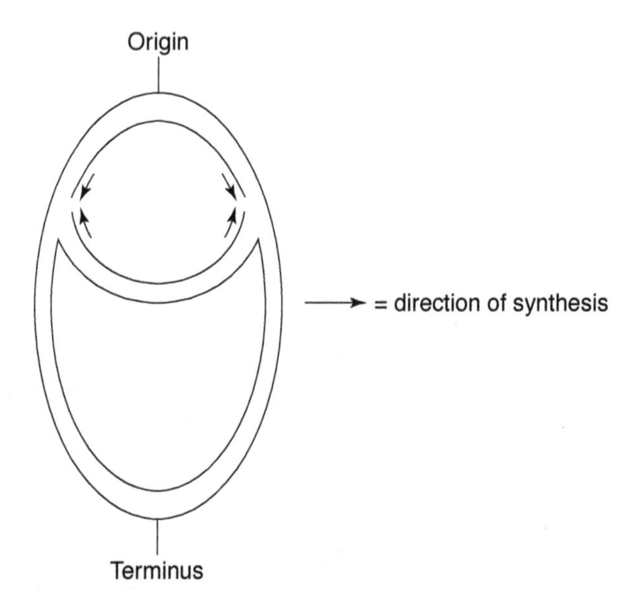

Initiation

At an origin, the replication process first involves DNA strand opening so that each strand of the DNA molecule is available as a template. Initiation is the rate-limiting step for replication of the chromosome. Like other metabolic pathways, the control of replication is exerted at the first committed step.

Initiation sequences contain a set of repeated sequences, which bind the essential initiator protein, DnaA. The DnaA protein opens the helix to make a short region of separated strands. Then a specialized single-strand binding protein binds to the DNA strands to keep them apart. This process makes a template, but replication can't happen because no primer yet exists. See Figure 8-10b.

Figure 8-10b

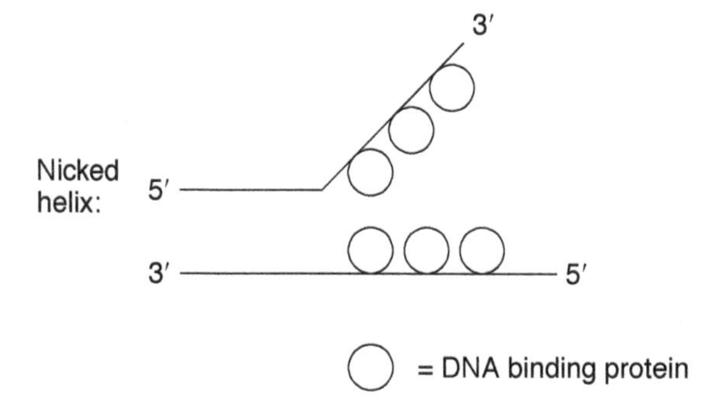

Chain initiation occurs when a specialized RNA polymerase enzyme called **primase** makes a short **RNA primer**. DNA polymerase III extends this RNA primer on both strands. Because DNA polymerase synthesizes DNA only in one direction (5' to 3'), only one strand is copied in each direction (left and rightward in the next figure). At the end of the initiation process, two **replication forks** exist, going in opposite directions from the "bubble" at the origin of replication, as shown in Figure 8-11.

Figure 8-11

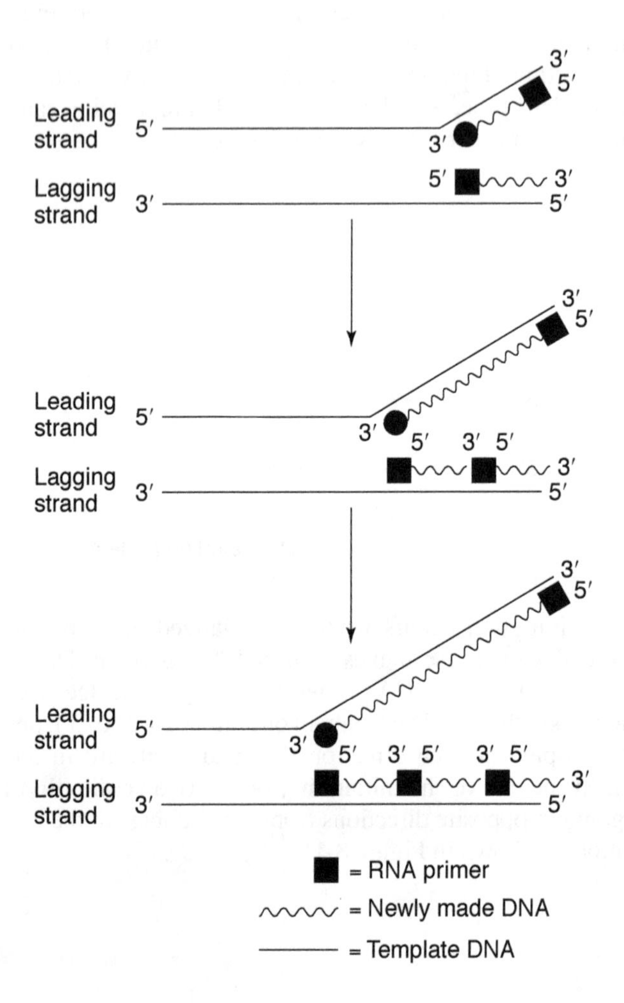
Leading strand
Lagging strand
5′
3′
= RNA primer
= Newly made DNA
= Template DNA

Elongation

Because only one strand can serve as a template for synthesis in the 5' to 3' direction (the template goes in the 3' to 5' direction, because the double helix is antiparallel), only one strand, the **leading strand**, can be elongated continuously. Ahead of the replication fork, DNA gyrase **(topoisomerase II)** helps unwind the DNA double helix and keep the double strands from tangling during replication.

Synthesis of the second (lagging) strand is more complicated because it is going in the wrong direction to serve as a template. No DNA polymerase exists to synthesize DNA in the 3' to 5' direction, so copying of the lagging strand is **discontinuous**—that is, short strands of DNA are made and subsequently matured by joining them together. An RNA primer, which is made by primase, initiates each of these small pieces of DNA. Then DNA polymerase III elongates the primer until it butts up against the 5' end of the next primer molecule.

DNA polymerase I then uses its polymerizing and 5' to 3' exonuclease activities to remove the RNA primer and fill in this sequence with DNA. Because Pol I is not very processive, it falls off the lagging strand after a relatively short-length synthesis. DNA polymerases can't seal up the nicks that result from the replacement of RNA primers with DNA. Instead, another enzyme, **DNA ligase**, seals off the nicks by using high energy phosphodiester bonds in ATP or NAD to join a free 3' hydroxyl with an adjacent 5' phosphate.

A multienzyme complex simultaneously carries out both leading and lagging strand replication. You can see the best model of the process in the next figure; the lagging strand may curl around so it presents the correct face to the enzyme. The two replication forks proceed around the chromosome, until they meet at the terminus. Termination is poorly defined biochemically, but it is known to require some form of DNA gyrase activity. See Figure 8-12.

Figure 8-12

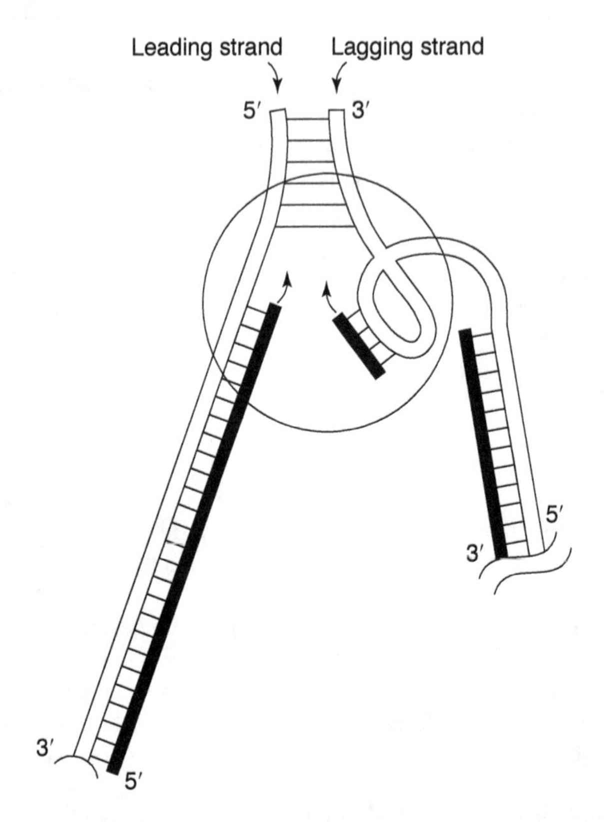

DNA Repair

DNA information is subject to loss or degradation. After all, DNA is a chemical, and chemicals can undergo reactions. Damage to a DNA leads to a loss or alteration of its information content, a process known as **mutation**. Compounds or agents that cause mutations are called **mutagens**.

Several types of mutations exist, depending on the changes to the sequence information of the chromosome. **Transition** mutations are defined as the conversion of G-C to A-T or A-T to G-C base pairs. For example, the compound hydroxylamine causes the loss of the amino group of deoxycytosine in DNA, converting it to deoxyuracil. When replicated, deoxyuracil is a template for deoxy-A incorporation. Hydroxylamine is a monodirectional mutagen, only converting G-C → A-T base pairs.

Transversion mutations convert purines to pyrimidines or vice-versa. For example, conversion of an A-T base pair to a T-A or C-G base pair is a transversion. Ultraviolet light can cause transversions, although not exclusively. **Frameshift** mutations result from the insertion or deletion of a single base pair. These are often due to planar intercalating mutagens, which insert into the double helix. More extensive deletions and insertions are common events, as shown in Figure 8-13.

Figure 8-13

- Transition: AT → GC or GC → AT
- Transversions: AT → TA or CG; GC → TA or CG
  (A pyrimidine is substituted by a purine and vice-versa.)
- Frameshift: XYZ → XWYZ(+1) or → XZ(−1)

Another important mutagen is ultraviolet light. Recent concern about the depletion of the atmospheric ozone layer by chlorofluorocarbon compounds (CFCs) is due to the role of the ozone in absorbing UV radiation before it can cause mutations in the organisms at the earth's surface. All the DNA bases efficiently absorb UV and become chemically reactive as a result. The formation of **pyrimidine dimers** from adjacent thymidine residues in DNA interferes with replication and transcription of DNA. See Figure 8-14.

Figure 8-14

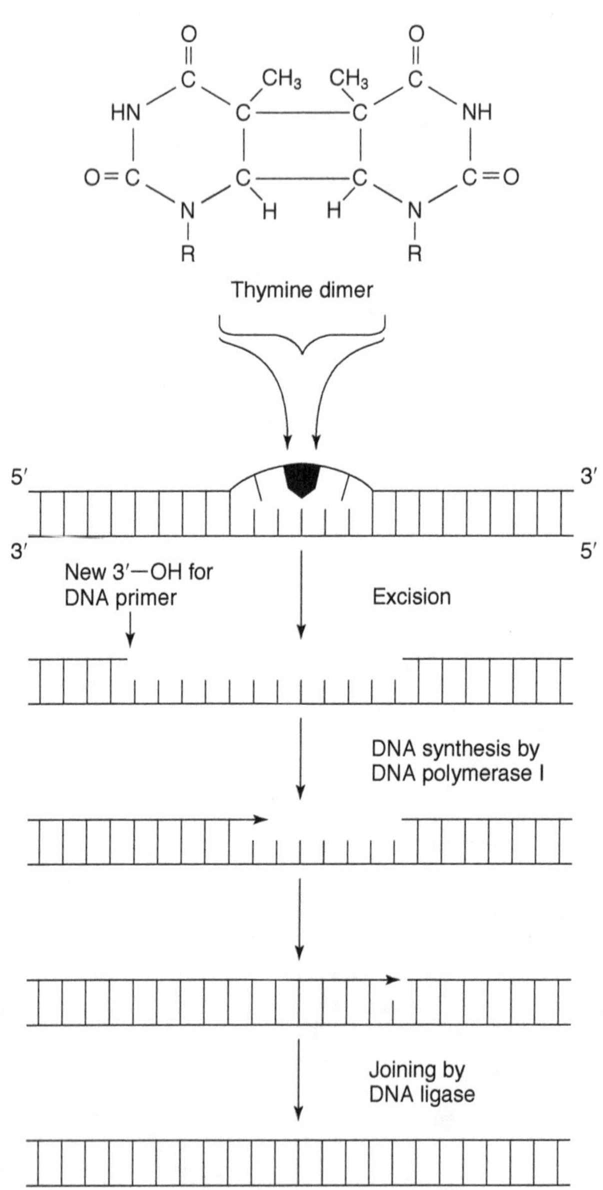
O
O
C
CH3
CH3
C
HN
C
C
NH
O=C
C
C
C=O
N
H
H
N
R
R
Thymine dimer
5′
3′
3′
5′
New 3′—OH for
DNA primer
Excision
DNA synthesis by
DNA polymerase I
Joining by
DNA ligase

Mutations can also be "spontaneous," due to errors in replication, background chemical reaction, and so forth. Because the rates of most chemical reactions increase with increasing temperature, and because mutagenesis is a chemical reaction, not surprisingly, even heat can be thought of as a mutagen.

DNA repair systems

Repair systems in the cell counteract mutations. If no way existed to prevent or revert mutations, life would be too unstable for one generation to give rise to another. Usually, the spontaneous mutation rate of an organism is less than one mutation per billion or so nucleotides per generation. Because humans have about three billion bases of DNA, even identical twins differ by several mutations. Viruses usually don't have DNA repair systems; they keep going from one generation to another either by inserting into a host's DNA or by releasing so many viruses per generation that enough good copies survive to select the active ones at the next generation.

A large number of systems counter mutagenic events. Direct reversal of the mutagenic event is possible in some cases. For example, methylation of guanine by certain DNA alkylating compounds leads to mutation. This form of DNA damage is counteracted by the synthesis of specific proteins that accept the methyl group from the damaged DNA and then are themselves degraded.

Ultraviolet light causes **photodimerization** of adjacent pyrimidine residues in DNA, such as the formation of thymidine-thymidine dimers.

Photoreaction can remove thymidine-thymidine photodimers. Visible light activates the enzyme **DNA photolyase**. This enzyme absorbs visible light and transfers that energy to the photodimer, reversing the dimerization directly.

Excision repair is a more general mode of counteracting mutagenic events. Many DNA alterations, including photodimerization, are reversed by cutting out the altered base and its neighbors, followed by synthesis of new DNA by DNA polymerase. In these processes, the double helix provides the information pointing to the "correct" DNA sequence. The initial event is the recognition of the altered DNA segment. For example, formation of pyrimidine dimers changes the shape of the double helix. Repair systems recognize the altered shape to start the repair process. After the repair systems detect the change, nuclease action near, but not necassarily at, the site of mutation cleaves the damaged chain. The cleaved segment is unwound extended, leaving a single-stranded gap. Then DNA polymerase I fills in the gap and DNA ligase closes the double helix.

Special systems remove deoxyuridine from DNA, as shown in Figure 8-15.

Sources of deoxyuridine in DNA include the presence of dUTP because dUMP (the substrate for thymidylate synthase) or dUDP (from ribonucleotide reductase action on UDP) are phosphorylated to the triphosphate. DNA polymerase recognizes these compounds as substrates. Another source is the deamination of deoxycytidine in DNA, promoted by a variety of compounds. If deoxyuridine is on a template strand of the DNA, it will direct the incorporation of an A in the newly made strand of DNA. This will convert a G-C pair to an A-T pair.

Cells have evolved ways of preventing mutations caused by either mechanism. First, cells contain an enzyme, **dUTPase**, which hydrolyses dUTP to dUMP. Thus the triphosphate substrate is taken away from DNA polymerase before it can serve as a substrate. Secondly, U residues in DNA from whatever source cause the gap repair system to be activated so that these alterations are removed before they can be replicated. The base-pairing in DNA of dU and dT are identical; each forms two hydrogen bonds with dA on the opposite strand. However, dT is used in DNA; this base contains a methyl group that dU and dC do not. This means that any time dU arises from the loss of the amino group of dC, it can be recognized and removed. If the base 5-methyl C were to be deaminated, that would lead to a T in the DNA. In this case, the repair system couldn't determine which strand of the double helix was incorrect; it would be likely to delete either the altered base or its partner, creating a high probability of mutation at that position. In fact, DNA is often methylated at some sequences—these sites on DNA are often mutational **hot spots**.

DNA Recombination and Repair

In cases where DNA is severely damaged, a cell will engage in a phenomenon called the **SOS response** in an effort to salvage a functioning set of genetic information. This response, also called **error-prone repair**, represents a last-ditch response to salvage a chromosomal information system. In addition, recombinational repair systems act to allow one copy of the replicating DNA at a replication fork to supply information to the other daughter chromosome. Recombinational repair is a way of using one copy of the cell's information to ensure that the overall information store remains intact.

The biochemical process of recombination occurs by breaking and rejoining DNA strands. The key reaction is strand displacement initiated at a nick in the chromosome. Then a protein called RecA (which stands for recombination; *rec*⁻ bacteria are unable to recombine their DNA information and therefore are abnormally sensitive to UV radiation) binds to a single-stranded DNA fragment and catalyzer its exchange with the same sequence of the duplex. RecA protein is a **strand displacement** protein. See Figure 8-16.

Figure 8-15

A G A

T U T

Uracil-DNA glycosidase

A G A

T T

AP exonuclease

A G A

T T

DNA polymerase I
DNA ligase

A G A

T C T

Figure 8-16

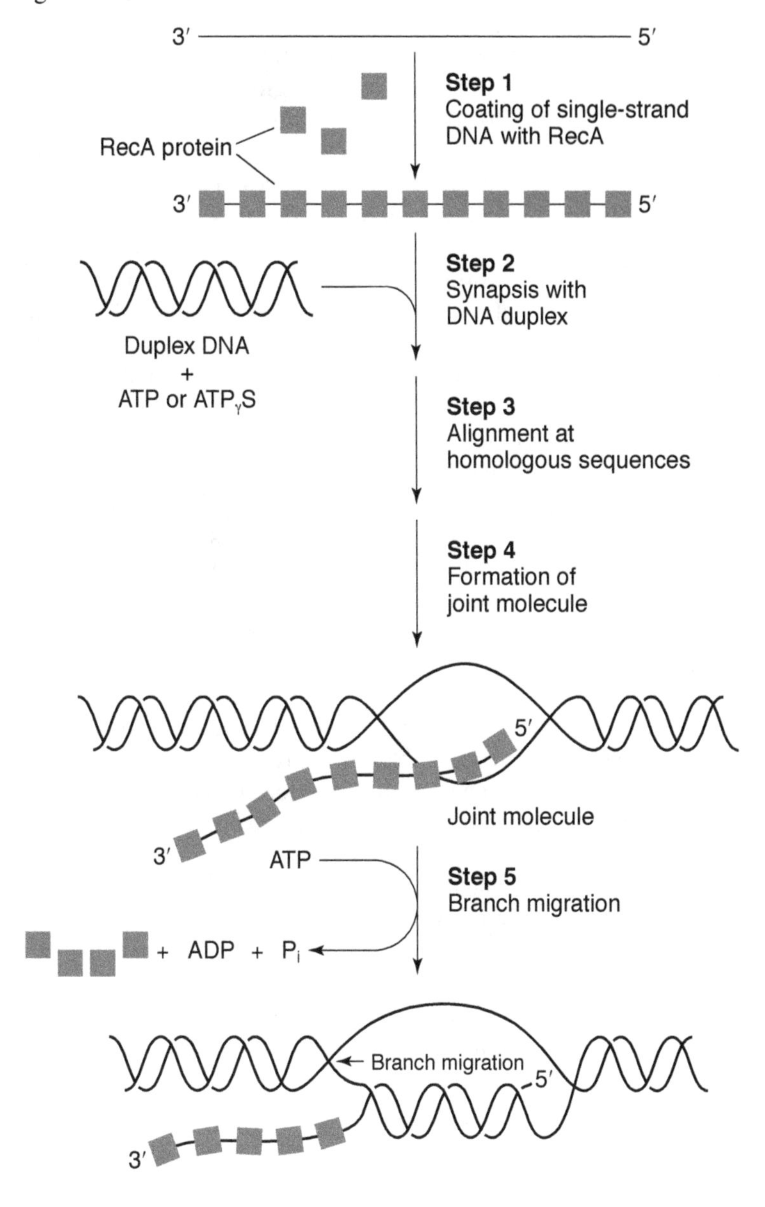
3′
5′
Step 1
Coating of single-strand DNA with RecA
RecA protein
3′
5′
Step 2
Synapsis with DNA duplex
Duplex DNA
+
ATP or ATPγS
Step 3
Alignment at homologous sequences
Step 4
Formation of joint molecule
5′
Joint molecule
3′
ATP
Step 5
Branch migration
+ ADP + Pi
Branch migration
5′
3′

RecA preferentially binds to single-stranded DNA in a **cooperative** fashion; this cooperativity means that RecA will cover an entire single-stranded DNA molecule rather than bind to several molecules partially. Rec A then aligns **homologous segments** (those with complementary information) to form base pairs. The key reaction of RecA-coated DNA is the movement of the single-stranded regions of the DNA to form a joint molecule—a process called **strand displacement**. This reaction involves ATP hydrolysis.

In **homologous recombination**, two double helices align and are nicked. Then RecA catalyzes the invasion of each double helix by one strand of the other. This forms a crossed structure called a **Holliday junction**. If the Holliday structure were simply broken at the point where it was formed, no genetic recombination could occur because the two original DNA molecules would simply reform. Instead, the junction **migrates** by displacement of one strand of DNA. Finally, the displaced Holliday junction is broken and rejoined, or resolved. The exact type of recombination between the two strands depends on which of the strands is broken and rejoined. Note that each recombination event involves two breaking and rejoining events: one to initiate strand displacement and one to resolve the Holliday junction. See Figure 8-17.

If the two DNAs have the same sequence, they can form a Holliday junction, but no detectable genetic recombination takes place because no information change has occurred. If the two DNAs are very different, no recombination will take place because formation of a Holliday junction requires homologous information. If the two DNAs of the Holliday junction are similar to each other but not identical (that is, they contain **mismatches**), then repair enzymes, which remove the base and/or nucleotide from one of the mismatched strands, will repair the DNA. The fact that some enzymes participate both in repair and in recombination accounts for the fact that many recombination-deficient mutant bacteria are also highly sensitive to ultraviolet light.

Figure 8-17

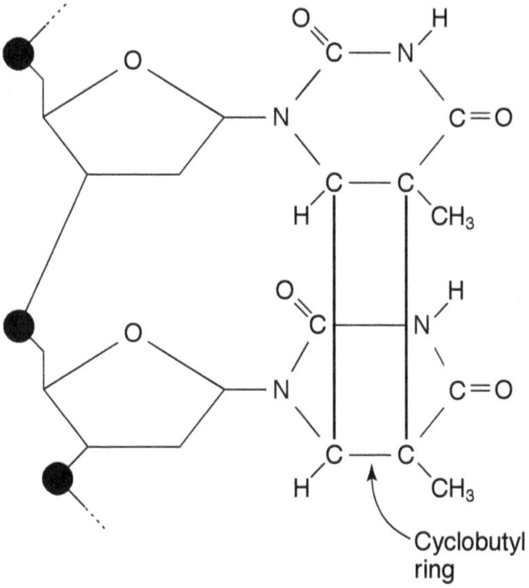

Cyclobutyl ring

The rare human genetic disease xeroderma pigmentosum is due to a deficiency in one of the many components of the DNA repair system. Exposure to ultraviolet light causes skin tumors. Individuals with this disease are so sensitive to ultraviolet light that they must avoid even household fluorescent lamps.

DNA STRUCTURE, REPLICATION, AND REPAIR

CHAPTER 9
MOLECULAR CLONING OF DNA

Gene Expression

With the knowledge of how to express genetic information comes the ability to alter that expression for useful purposes. In one sense, humans have done this since the agricultural revolution. For example, early North Americans learned how to breed varieties of the grass teosinte so that the offspring would produce seeds that were less hard and simultaneously more plentiful. The result was maize. More selective breeding for yield and disease resistance has led to hybrid corn varieties today. Similarly, humans has bred animals for desirable properties, such as horses for speed or power, or dogs for gentle temperament, strength, and so on. DNA-based genetics is a continuation of that same sort of breeding with two differences: first, the DNA is manipulated biochemically rather than in a genetic mating, and second, DNA can be exchanged between different species (which happens only rarely in nature, although it does occur).

DNA and Information

Whether you call the field of study biotechnology, genetic engineering, or DNA technology, the goal of DNA-based genetics is to analyze and manipulate DNA on the basis of its *information* content. In other words, this field studies the *sequence* of the individual nucleotides in a particular molecule. This task can be difficult, because one molecule of DNA looks much like any other, which implies that genes can't be separated by ordinary chemical methods. The DNA of bakers' yeast looks much like the DNA of the organism that produces penicillin, yet the organisms whose properties the DNA encodes are very different. Chemically, DNA maintains a very regular structure; differences in base composition are less important than sequence differences in determining the information content of DNA. For example, if one were to make a tetranucleotide (a DNA four units long) from one molecule each of A, G, C and T, 4! (4 × 3 × 2 × 1) possible arrangements would exist—AGCT, AGTC, ACGT, ACTG, ATCG, ATGC, GCTA, GTCA, GATC, and so on. However, all of the molecules would have the same base composition—an equal amount of A, G, C, and T. Ordinary means of organic chemistry would be poor at separating such closely related molecules, and the problem would get much harder as the DNA molecules got bigger.

The question of scale makes the problem more difficult. The human genome contains about 3 billion base pairs of DNA. Therefore, each individual gene makes up only a very small fraction of the total amount of DNA. Moreover, not all DNA in the cell functions. Complex organisms (multicellular ones such as humans, for example) carry a large load of noninformational DNA as a part of the total DNA of a cell. As a result, before the advent of DNA cloning techniques, obtaining enough of any particular DNA sequence to study in detail was very difficult. DNA technology depends on the ability to separate nucleic acids on the basis of their sequences, that is, their information content.

DNA Hybridization

Nucleic acid hybridization allows scientists to compare and analyze DNA and RNA molecules of identical or related sequences. In a hybridization experiment, the experimenter allows DNA or RNA strands to form Watson-Crick (see Chapter 4 of Volume I) base pairs. Sequences that are closely related form base-paired double helices readily; they are said to be **complementary**. The amount of sequence complementarity is a measure of how closely the information of two nucleic acids relate. The complementary strands can be both DNAs, both RNAs, or one of each.

Heating the DNA solution above a characteristic temperature can separate the two strands of a double helix. That temperature is called the **melting temperature**, abbreviated $\mathbf{T_m}$. Above the T_m, a DNA is mostly or all single-stranded; below the T_m, it is mostly double-stranded. For a natural DNA, the T_m depends primarily on its G+C content. Because a G–C base pair has three hydrogen bonds and an A–T pair only has two, nucleic acid double helices with a high G+C content have a higher T_m than do those with a greater proportion of A+T. The T_m is not an exact property: It depends on the solvent conditions. For example, a high concentration of salt (such as NaCl) raises the T_m of a DNA duplex, because the positive Na^+ ions shield the negative charges on the phosphodiester backbone from repelling each other. Likewise, certain organic solvents can cause the negative charges on the phosphates to repel more strongly; these solvents lower the T_m of a DNA double helix.

What happens if two nucleic acids are partly complementary and partly different? In this case, some stretches of the two strands may form base pairs while others don't. The two molecules can be manipulated so that they form a hybrid or separate. The conditions favoring the formation of duplex nucleic acid are low temperature (below the T_m), high salt, and the absence of organic solvents. The latter two conditions raise the T_m of the hybrid duplex so that the DNA would remain more double-stranded. On the other hand, higher temperatures (closer to the T_m of the hybrid) lower salt, and the presence of organic solvents would tend to push the two strands of the DNA apart.

The term **stringency** sums up these variables: The more stringent the conditions, the more likely partially complementary sequences are to be forced apart. Conversely, less stringent hybridization conditions mean that the two strands need not be so complementary to form a stable helix. See Figure 9-1.

Figure 9-1

High stringency (low salt, high temperature): Associated DNAs can have few mismatches.

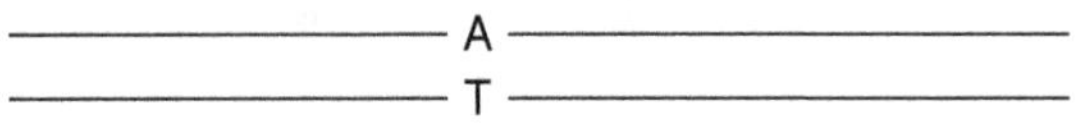

———————— A ————————
———————— T ————————

"Matched" pair

Low stringency (high salt, low temperature): Associated DNAs can have more mismatches.

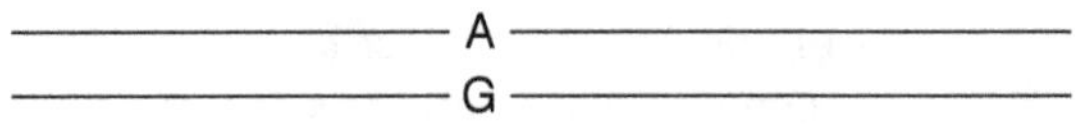

———————— A ————————
———————— G ————————

"Mismatched" pair

Hybridization can be used to classify the DNAs of various organisms. For example, human DNA is 98 percent identical to that of chimpanzees, and these two DNAs form a duplex under stringent conditions. Related sequences of humans and birds can also form hybrids, but only at a much lower stringency.

Restriction Enzyme Mapping

Restriction enzymes function like a primitive immune system. Bacteria use these enzymes to cut DNA from foreign sources, like the viruses that infect them (called bacteriophage). The cut DNA can't be used to make new bacteriophage, so while the virus may kill one bacterial cell, the culture of bacteria as a whole will be spared. Bacteria refrain from cutting their own DNA by modifying it, usually by methylating it, so that it is not a substrate for the restriction enzymes.

Molecular biologists use restriction endonucleases (the term **endonuclease** means that the enzyme cuts nucleic acids in the middle of a molecule rather than from one end) to cleave DNAs because these enzymes cut only DNA, and more importantly, only at particular short sequences in the double helix. The most useful enzymes are the so-called Type II restriction enzymes that cut double-stranded DNA at short palindromic sequences (**palindromes** are sequences that read the same in the 5' to 3' direction). For example, the enzyme EcoRI (from certain strains of the bacterium *E. coli*) cleaves DNAs at the sequence 5' GAATTC 3', which is the same when read on the opposite strand. EcoRI endonuclease cuts this sequence on both strands, between the G and the first A (the phosphate of the phosphodiester bond goes to the 5' position of the A). After the two pieces of DNA separate, the 5' pAATT portion of each piece is single-stranded. These single-stranded tails are called **cohesive ends**. Other Type II restriction enzymes generate **blunt ends** by cutting in the middle of the palindrome. For example, the enzyme SmaI (from the bacterium *Serratia marcescans*) cuts DNA in the middle of the sequence 5' CCCGGG leaving two strands with blunt ends.

Figure 9-2

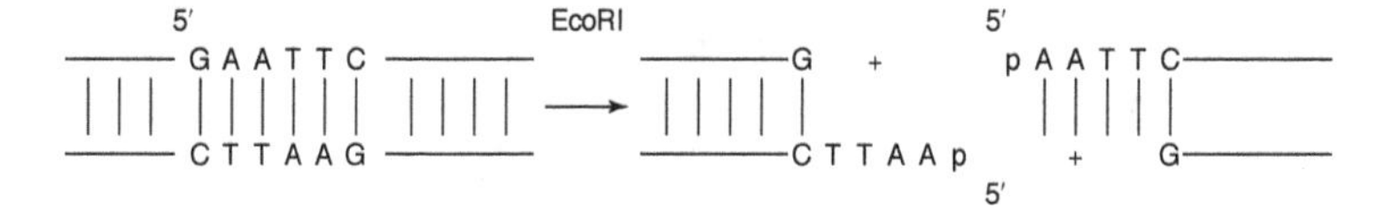

As mentioned previously, bacteria defend themselves from their own restriction enzymes by methylating their DNA. For example, the EcoRI methylase adds a single methyl group to the amino group of the second adenine of the recognition sequence. Methyladenine at this site on either one of the DNA strands is sufficient to prevent the cleavage of the DNA at this site by the EcoRI endonuclease.

DNA Mapping

Recognition sequences for restriction enzymes are like signposts on a DNA molecule. Just as a highway map gives the important features along a road, restriction enzymes can give a map of features along a DNA molecule. Thus, for example, the circular DNA molecule of a small virus can be completely digested by a restriction enzyme to make a so-called **limit digest**. Putting the fragments into a gel and passing electrical current through the gel separates the fragments. Because DNA has negatively charged phosphates at each phosphodiester bond, the positive direction of the gel will attract the DNA. The DNA fragments start to move, but the larger the fragment, the more often it "bumps into" the gel. As a result, the DNA fragments are separated by size, with the smallest fragments moving farther in the gel, as shown in Figure 9-3.

After the sizes of the fragments are known, they can often be ordered by several strategies. The simplest (in theory if not in practice) is to determine the size and composition of *partial digestion products* that contain two or more of the limit fragments. For example, if a DNA contains EcoRI fragments of 1500, 2000, and 3000 base pairs, a partial digestion that gives fragments of 3500 and 4500 base pairs must mean that the order of fragments is either 2000-1500-3000 or 3000-1500-2000. If the DNA is circular, the two maps are interchangeable.

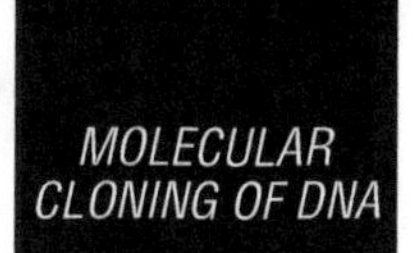

Figure 9-3

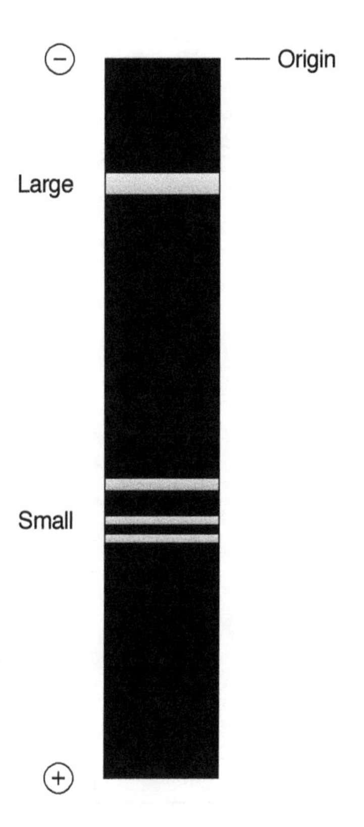

Gel Electrophoresis of DNA

Figure 9-4

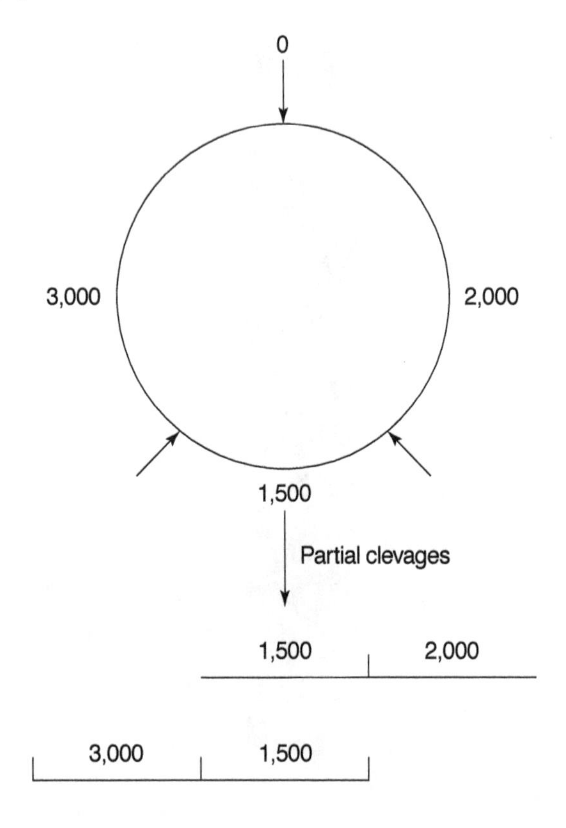

Southern Blotting

This technique, named after its inventor E.M. Southern, is a way of combining restriction enzyme and hybridization information. In Southern blotting, a DNA is digested with a restriction enzyme and the fragments are separated by gel electrophoresis. The DNA fragments are denatured by soaking the gel in base and then are transferred to a piece of nitrocellulose filter paper by capillary action. The result is like a "contact print" of the gel; each fragment is at a position on the paper that corresponds exactly to its position in the gel. Then the filter paper is mixed with a radioactively labeled single-stranded "probe" DNA or RNA. The probe DNA hybridizes to the complementary single-stranded DNA fragments on the filter paper. The blot is exposed to X-ray film and the resulting **autoradiograph** shows the positions of the complementary fragments of DNA. This information physically maps the regions of a large DNA that are complementary to a probe. See Figure 9-5.

Figure 9-5

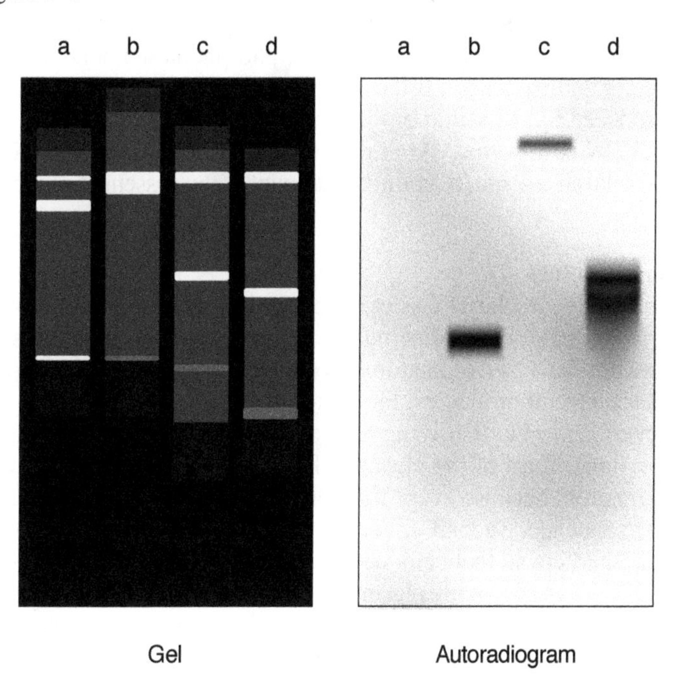

Molecular Cloning

DNA replication involves the copying of each strand of the double helix to give a pair of daughter strands. Replication begins at a specific sequence, called the **origin**. After initiation begins at an origin sequence, all sequences are replicated no matter what their information. This principle leads to the idea of molecular cloning, or recombinant DNA. Cloning enables the production of a single DNA sequence in large quantities.

A recombinant DNA consists of two parts: a **vector** and the **passenger** sequences. Vectors supply replication functions—the origin sequences to the recombinant DNA molecule. After it joins to a vector, any passenger sequence can be replicated. The process of joining the vector and passenger DNAs is called **ligation**. **DNA ligase** carries out ligation by using ATP energy to make the phosphodiester bond between the vector and passenger. If the vector and passenger DNA fragments are produced by the action of the same restriction endonuclease, they will join by base-pairing. After they are ligated to a vector, it is possible to make an essentially unlimited amount of the passenger sequence.

Plasmid vectors

Plasmids are circular DNAs that are capable of independent replication. Many naturally occurring bacteria contain plasmids; plasmid vectors are derived from naturally occurring plasmids. Plasmid vectors have several properties. First, they contain single restriction sites for several enzymes. Cleaving with one of these enzymes generates a single, linear form of the plasmid. This feature helps to ensure efficient ligation, because every ligation product will contain the entire vector sequence. Secondly, vectors are made to contain selectable genetic markers so that cells which contain the vector can be propagated. For example, many plasmid-cloning vectors contain a gene that encodes resistance to the antibiotic ampicillin (a member of the penicillin family). Cells that contain the vector can then be selected for by growth in the presence of the antibiotic; cells that don't contain the vector sequence will be inhibited by ampicillin in the growth medium, as shown in Figure 9-6.

Figure 9-6

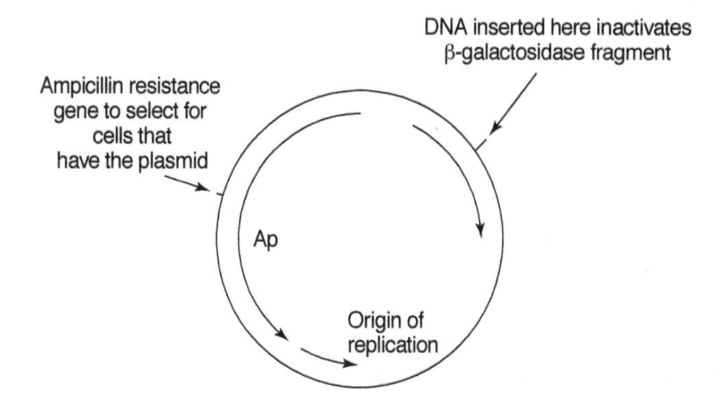
PLASMID VECTOR
DNA inserted here inactivates β-galactosidase fragment
Ampicillin resistance gene to select for cells that have the plasmid
Ap
Origin of replication

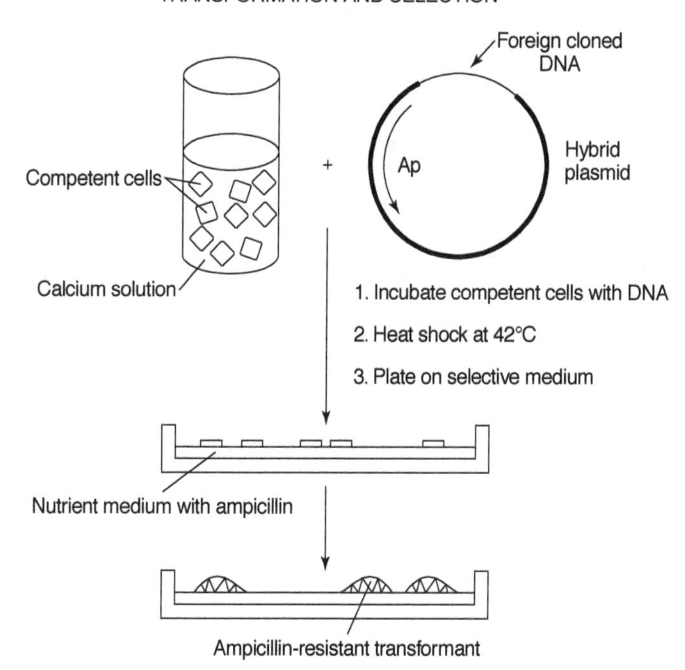
TRANSFORMATION AND SELECTION
Foreign cloned DNA
Competent cells
+
Ap
Hybrid plasmid
Calcium solution
1. Incubate competent cells with DNA
2. Heat shock at 42°C
3. Plate on selective medium
Nutrient medium with ampicillin
Ampicillin-resistant transformant

Thirdly, a means to detect which cells have only the plasmid vector as opposed to the recombinant product must exist. This determination is usually accomplished by a mechanism called **insertional inactivation**. The idea of insertional inactivation is that inserting passenger DNA into a gene interrupts the sequence of the gene, thereby inactivating it. For example, the restriction sites of many common plasmid-cloning vectors are located in a fragment of a gene for β-galactosidase, an enzyme involved in lactose metabolism. When cells containing just the vector are grown in the presence of an artificial substrate related to lactose, the colonies turn blue, because active enzyme is made. On the other hand, when the restriction site has a piece of foreign DNA inserted into it, the gene cannot make an active protein fragment because the DNA sequence interrupts the coding sequence of the gene. As a result, colonies of the bacteria that contain cloned foreign DNA appear whitish. The bacteria that are present in the colony can be grown separately, and standard biochemical procedures easily isolate the recombinant DNA they contain. If no such selection existed, each colony would have to be grown separately and its DNA analyzed—a very "hit or miss" proposition in many cases.

After a recombinant plasmid has been formed in the laboratory, it must be replicated. This process begins with the growth of the cell containing the recombinant plasmid. Plasmids are usually transferred to new hosts by **transformation**. Transformation is the addition of naked foreign DNA into a recipient bacterium. Because the plasmid usually contains a gene encoding the ability to grow in the presence of an antibiotic, identifying a recombinant bacterium is a two-part process. First, growth in the presence of the antibiotic ensures that each bacterial colony contains a plasmid, while the color of the colony identifies the plasmids that contain inserted DNA. Finally, the growth of the transformed bacterial colony **amplifies** the clones, that is, makes more copies of the plasmid. These steps can be done with a known piece of DNA or with a less-defined collection of insert DNA—for example, all the DNA from an organism. A collection of different cloned DNAs is called a **library**. The number of independent sequences in a library is called its **complexity**; the more complex the library, the greater number of independent sequences it contains.

Other types of cloning vectors

Viruses that infect bacteria are called **bacteriophages**. Native bacteriophage have been formed into vectors that can also be used for cloning. Bacteriophage vectors have three advantages over plasmids. First, they can carry significantly more foreign DNA than can plasmids, which are limited to about 5,000 base pairs (5 kilobases) of foreign DNA. Plasmids with larger-sized inserts tend to be unstable when amplified. The larger the insert, the fewer independent clones are required to have a reasonable chance of identifying any single gene or sequence in the collection of cloned DNA. Secondly, the virus particles of bacteriophage vectors can accept DNA only of a narrow size range. This means that DNA can be preselected so that each recombinant virus will contain only a single foreign sequence. This property is a consequence of the fact that DNA (whether native or foreign) must be packaged into a protein coat to be infectious. DNAs that are too large or too small can't be packaged efficiently, so they will not be represented in the library. Finally, bacteriophage infection can be a very efficient process, with nearly 100 percent of the packaged virus particles leading to a productive infection. In contrast, cells take up only one of 100,000 plasmid molecules in a transformation procedure. Bacteriophage clones are amplified by repeated cycles of infection and growth. See Figure 9-7.

Figure 9-7

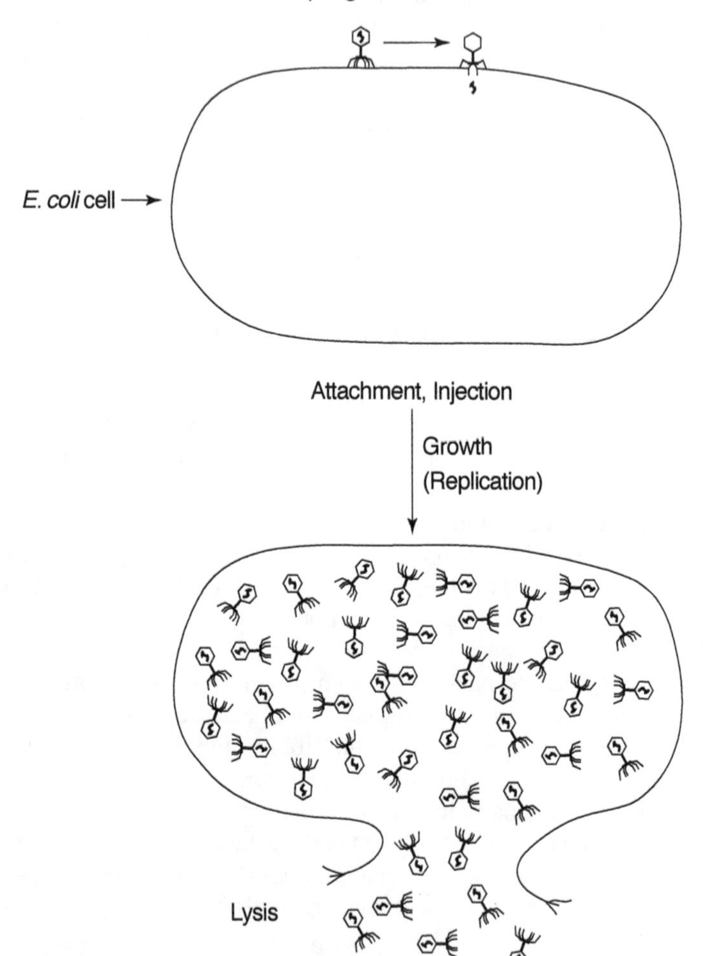
Bacteriophage with recombinant DNA
E. coli cell
Attachment, Injection
Growth
(Replication)
Lysis
Release of 100-150 Phage

Identifying Particular Sequences in a Library

Two types of libraries exist in an organism, depending on the source of the DNA used to make the library. Cloning total DNA from the cell of an organism makes a genomic library. Genomic libraries thus contain all types of sequences, including those which never find their way into messenger RNA (for example, the promoters of genes, or especially, the introns that are found in some or all genes of an organism). On the other hand, cDNA libraries (c stands for copy) are made by first converting mRNA into a DNA copy, a process known as **reverse transcription**. Then the copy DNA (cDNA) is cloned into a plasmid or bacteriophage vector. Clearly, the probability of isolating a desired DNA sequence in either a cDNA or genomic library depends on the complexity of the DNA source and the number of independent clones.

Specific clones are screened for or selected from the recombinant library. It is relatively simple to see how **selection** can be done if the desired DNA clone contains a gene required for growth of the host. For example, suppose one wanted to work with the gene sequence that specified an enzyme required for the biosynthesis of leucine. Bacteria that lacked that enzyme would not grow unless the media in which they were growing supplied leucine to the bacteria. If a plasmid library were transformed into those mutant bacteria and the transformants were plated on agar plates that lacked leucine, only the bacteria that contained the cloned DNA sequence encoding the missing enzyme could grow. This is an easy experiment to do but it doesn't always work. In general, the more distantly related the source of the DNA, the more likely that the cloned DNA can be expressed to make a functional product. For example, many other bacteria would contain the enzyme, and the cloned DNA would likely be expressed. On the other hand, DNA from a plant or animal source that encoded the enzyme would likely contain introns and not be expressed in the bacterium.

Hybridization **screening** uses a nucleic acid probe in an experimental setup very much like a Southern blot. Recombinant bacteria or bacteriophage are grown on a petri plate and then partially transferred to filter paper. The filter paper is then treated to fix the DNA in place, and a specific DNA fragment (or probe) is hybridized to the DNA on the filter paper. The radioactive or otherwise labeled DNA probe sticks to the filter paper only where it contains complementary sequences. Because the blot is like a contact print of the position of the colonies on the petri plate, the location of the complementary sequences serves as the key for amplifying the desired clones. Probes are often made from knowing a small region of the amino acid sequence of a purified protein, figuring out the possible sequences that can encode that amino acid sequence from the genetic code, and then chemically synthesizing all the DNAs that can encode that amino acid sequence. After a single clone is identified, its DNA can be used as a probe to find overlapping clones and the whole assemblage can be fitted into a map of the genc of interest, as shown in Figurc 9-8.

Figure 9-8

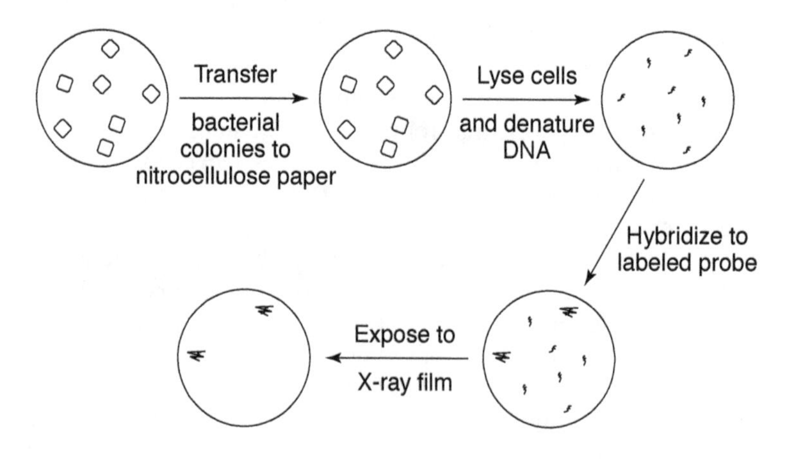

The Polymerase Chain Reaction

You can isolate virtually any DNA sequence by means of the **polymerase chain reaction**, or **PCR**. PCR uses repetitive cycles of primer-dependent polymerization to amplify a given DNA. Very little original DNA is required, as long as two unique primers are available. Knowing the sequences of the primers before starting out is helpful, but not always necessary. Each cycle of PCR involves three steps: DNA double strand separation, primer hybridization, and copying. First, the original DNA is denatured by heat treatment to make two separated strands. Then the two primers are hybridized to the DNA, one to each of the two separated strands. These primers act as initiators for DNA polymerase, which copies each strand of the original double-stranded DNA. The original two strands of DNA now become four strands, which are then denatured. These four strands are then hybridized with the primers and each of them is now copied, to make eight strands, and so forth. Amplified DNA can be analyzed by any of the techniques used for analyzing DNA: it can be separated by electrophoresis, Southern blotted, or cloned. Because a single DNA sequence is obtained by PCR, sequence information can also be obtained directly. See Figure 9-9.

Figure 9-9

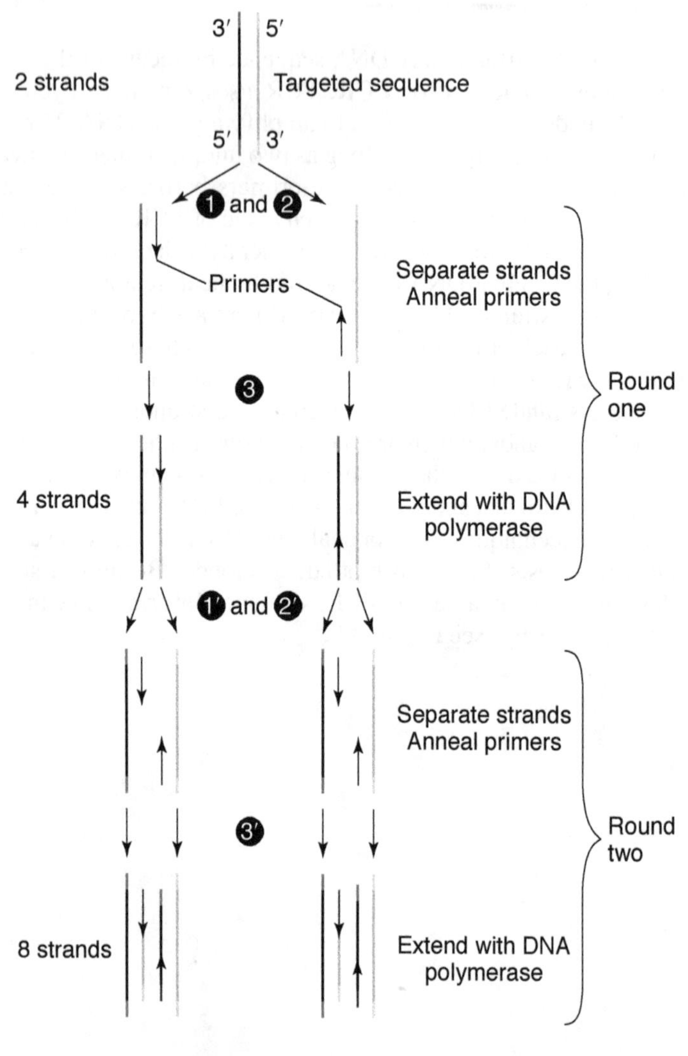

Two Cycles of PCR

DNA Sequence Determination

The most important information about DNA is its nucleotide sequence. DNA sequences are determined by exploiting the enzymatic characteristics of DNA polymerase. DNA polymerases all require a primer to initiate synthesis. The sequencing reaction begins when a single primer is hybridized to the strand of DNA to be sequenced. This primer-initiated synthesis is just like the initiation of a PCR. The primer used to initiate synthesis can be complementary to the vector in which the DNA is cloned, or it can be a primer used to amplify the DNA in PCR. The important point for sequencing DNA is that all of the newly synthesized molecules will have the same sequence. The primer is extended by addition of deoxynucleoside triphosphates along with DNA polymerase. If these were the only components of the reaction mixture, then all the newly extended molecules would be complementary to the entire template chain, just as in PCR. The sequencing reaction also contains, in addition to the four deoxynucleoside triphosphates (dNTPs), 2',3'-dideoxynucleoside triphosphates, or ddNTPs (ddGTP, ddATP, ddTTP, or ddCTP), as shown in Figure 9-10.

As the preceding figure shows, the ddNTPs lack both the 2' and 3' hydroxyl groups of a ribonucleoside triphosphate; therefore, if a ddNTP is incorporated into the growing DNA chain, all synthesis stops. Because a ddNTP must be base-paired for it to be inserted into a chain, the incorporation of a dideoxynucleotide at a position is an indicator of the nature of the complementary base on the chain. In other words, ddG is incorporated into the chain only when a C is present in the template, ddA only in response to a T, and so on. Because either a deoxy- or dideoxynucleotide may be inserted at a given site, the result is a set of nested chains, beginning at the primer and ending at a dideoxynucleotide. The members of this collection of molecules all have the same 5' end (because they all start with the same primer) and their 3' ends terminate with a dideoxynucleoside. If the molecules are separated according to chain length by electrophoresis through a thin polyacrylamide gel, then the nucleotide sequence of the corresponding DNA can be read directly. The DNA fragments are detected by fluorescence. Sequences are read automatically from a single lane because each terminator is modified to contain a different fluorescent dye, so that each dideoxynucleoside is associated with a single color of the fluorescence. By using these methods, it is possible to read 700 or more nucleotides in a single experiment.

Figure 9-10

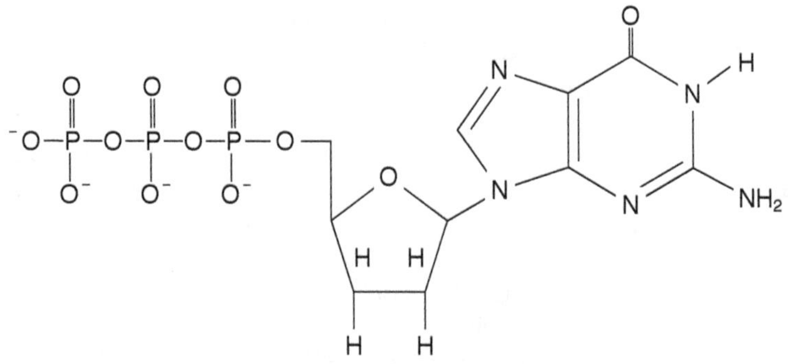

Dideoxy GTP (ddG)

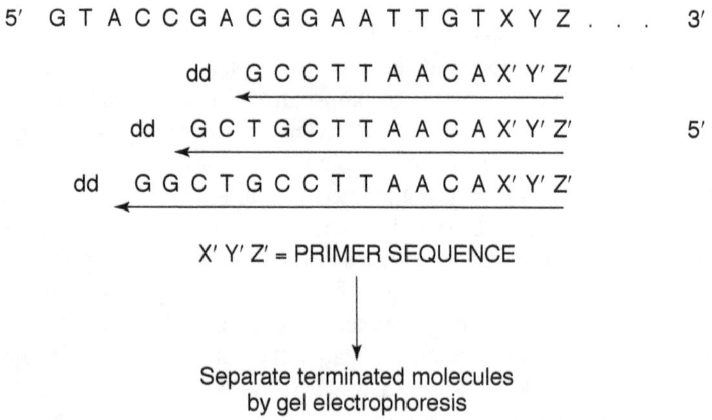

5′ G T A C C G A C G G A A T T G T X Y Z . . . 3′

dd G C C T T A A C A X′ Y′ Z′

dd G C T G C T T A A C A X′ Y′ Z′ 5′

dd G G C T G C C T T A A C A X′ Y′ Z′

X′ Y′ Z′ = PRIMER SEQUENCE

Separate terminated molecules
by gel electrophoresis

Genomics

With easy DNA sequencing technology comes the possibility to determine all the information present in the DNA of a single organism—the entire sequence of its DNA. **Genomics** is the study of an organism's entire information content. In humans, that amounts to about 3 billion base pairs of information, or about 10 million sequencing experiments (although it isn't nearly all that efficient). More to the point is the sequence information in a small bacterium, about 4 to 5 million base pairs of information. This is much more feasible, and can be accomplished in about 10,000 to 20,000 separate experiments—a few weeks' worth of information gathering in a big, dedicated laboratory. The strategy is sometimes called "shotgun sequencing," because the information-gathering process isn't precisely aimed. Rather, a random set of clones is made in a bacteriophage-like vector (called a **cosmid**) and a mini-library of DNA inserts is made in a plasmid vector. These recombinant plasmids are then used as templates for sequencing reactions such as the ones described earlier. The sequences are analyzed by computer so that the overlapping DNA sequences can be fit together to make longer sequences. This step-by-step process is continued until enough information is gathered to assemble a whole sequence of the DNA for an organism or chromosome. Although the shotgun technique may lead to a single stretch of sequence being determined several times before all the sequences are determined, it is faster than trying to order each clone and then determine their sequences individually.

Scientists have determined complete genomic sequences for a number of viruses, over fifteen bacteria, and laboratory yeast. Many pharmaceutical companies are interested in the sequence information from bacteria that cause disease, in the hopes that the information obtained will lead to new drug targets.

CHAPTER 10
RNA AND TRANSCRIPTION

RNA Information

The information in DNA remaining stable is essential. You can think of DNA as the "master copy" of a computer program. When you get a new program, you first copy it from the purchased disk onto the hard disk of the computer and use that copy as the source of the program for use. You store the original copy of the program away and only use it if the first one crashes. If DNA information were to be used regularly in the cell, it could accumulate errors, which would be passed on from one generation to the next. Before too long, the DNA would have so many errors that it would lose important functions and couldn't support the organism. Just as a clever computer user avoids this problem by making copies of the programs before using them, cells use copies of their genomic information for the working processes of information flow (essentially protein synthesis) in the cell. These copies are made of a **related nucleic acid**, RNA.

RNA molecules are structurally and metabolically distinct from DNA molecules. First of all, the sugar present on the nucleotide is ribose and not deoxyribose. This has the consequence that an RNA chain is less stable than the corresponding DNA because the 2'-OH group makes the 3',5' phosphodiester chain of RNA hydrolyzable in alkaline solution. In an alkaline solution, OH^- ions in the solution remove a proton from the 2'-OH of ribose; the 2'-O^- then is attracted to the central, relatively positive, phosphorus. The resulting intermediate can be resolved by the cleavage of the phosphodiester bond and the breaking off of the next nucleotide in the chain as shown in Figure 10-1.

Figure 10-1

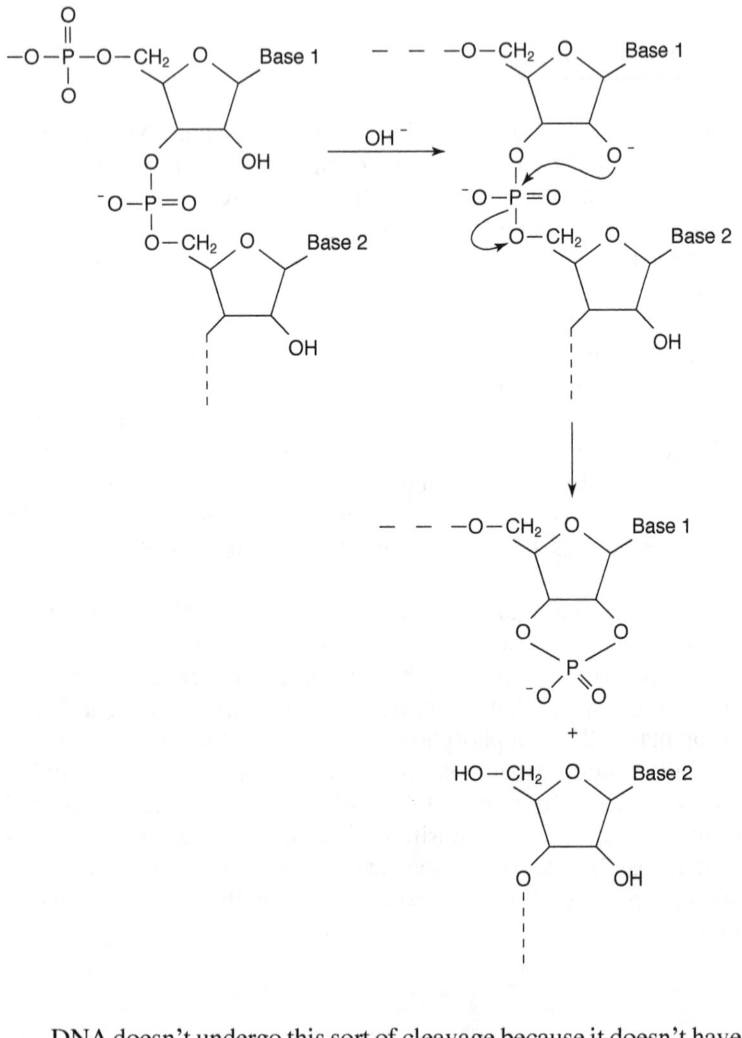

DNA doesn't undergo this sort of cleavage because it doesn't have a 2'-OH group to be ionized in the presence of hydroxide ion. As discussed previously, when DNA is exposed to base, the two strands of the double helix separate but otherwise aren't altered. Secondly, RNAs are almost always single-stranded. This means that RNA information can't be repaired, at least in the way that DNA is repaired (by the use of the information on one strand as a template to direct the synthesis of the complementary strand). Again, this property is consistent with the instability of RNA during cell metabolism. The fact that RNA is single-stranded means that **intramolecular** base-pairing determines its structure. This property gives RNA chains a compact secondary structure, in contrast to the extended structure of a DNA double helix. See Figure 10-2.

Figure 10-2

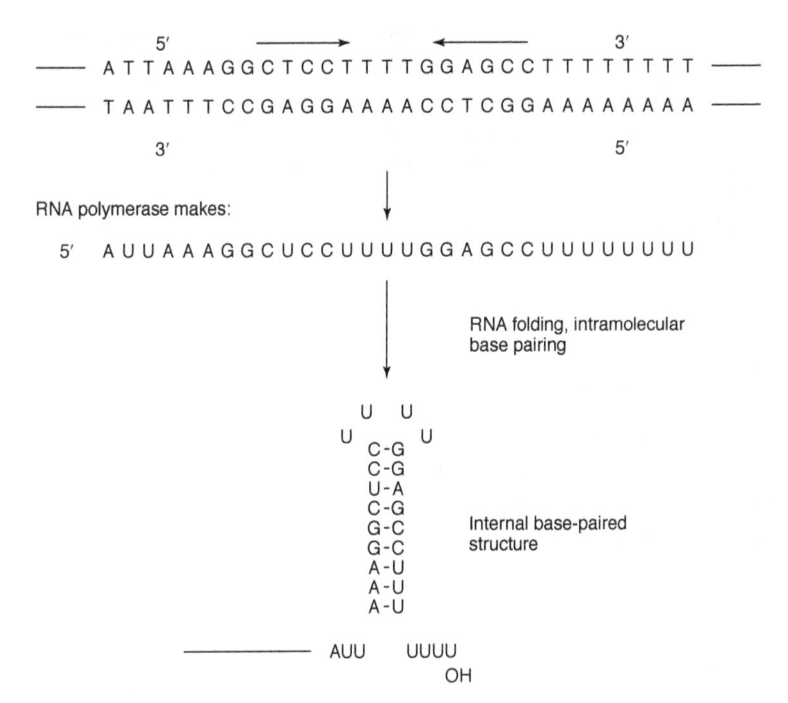

Not all the base-pairing is of the Watson-Crick variety, either. For example, G-U, U-U, and A-G "base pairs" are relatively common. These interactions contribute to the wide variety of structures that RNA can assume. RNA molecules also contain modified nucleosides, and in some cases, quite complicated ones. These are synthesized post-transcriptionally as part of the maturation process of RNA and serve to "fine-tune" RNA functions.

RNA Functions

Most RNAs function in an information carrying and/or processing mode in the cell. As Chapter 4 of Volume 1 of this series pointed out, the overall information processes of the cell are given in the Central Dogma of Molecular Biology: DNA makes RNA makes Protein. RNA is involved as a carrier of information, as translator or adaptor of RNA information into protein information, and as catalyst for the synthesis of the peptide bond.

Messenger RNA

Prokaryotic and eukaryotic messenger RNAs (mRNAs) have different structures. Prokaryotic mRNAs are often polycistronic (that is, they carry the information for more than one protein) while eukaryotic mRNAs are monocistronic and almost always code for a single protein. Eukaryotic mRNAs also have structural features that prokaryotic ones do not. While prokaryotic mRNAs generally have only the common four bases A, C, G, and U, eukaryotic mRNAs contain a modification known as a **cap** at the 5' end. The cap is a complex structure with an unusual 5'-5' arrangement **phosphodiester linkage**. In addition, some of the nucleosides in the cap are methylated: The most 5' base is 7-methylguanosine, and one or two other sugars are methylated at the 2' oxygen as shown in Figure 10-3.

Figure 10-3

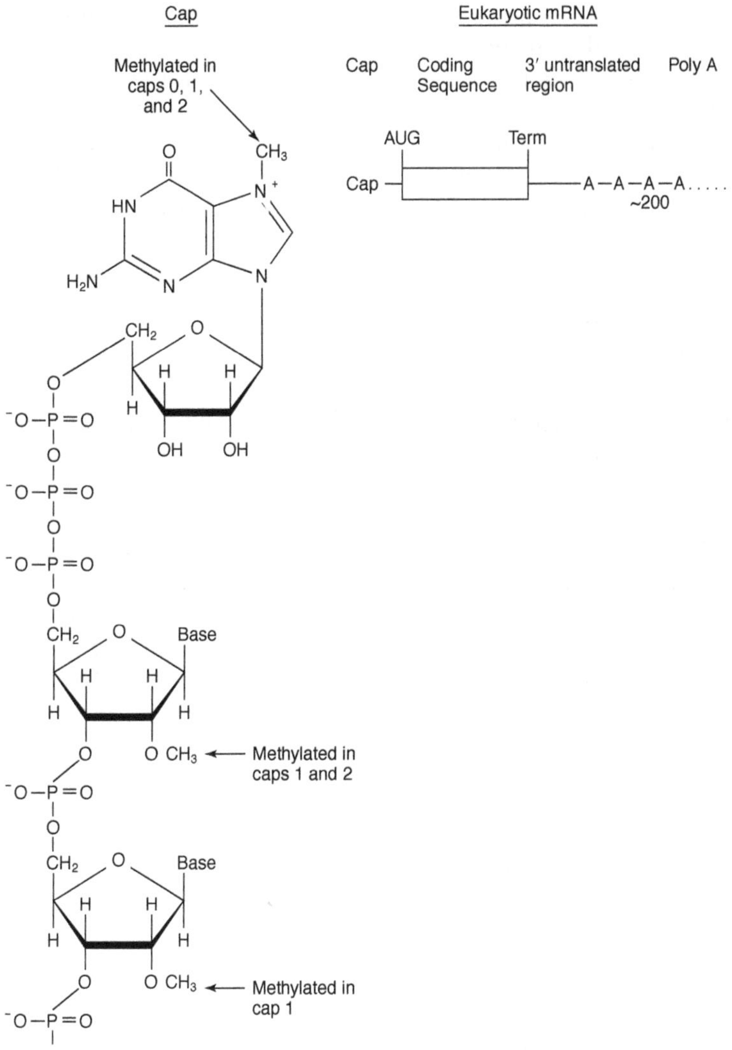

Cap

Methylated in caps 0, 1, and 2

CH_3

Methylated in caps 1 and 2

Methylated in cap 1

Eukaryotic mRNA

Cap | Coding Sequence | 3′ untranslated region | Poly A

AUG | Term

Cap — A—A—A—A.....

~200

Eukaryotic mRNAs often have long 3' untranslated sequences—sequences that follow the stop codon for the protein they encode. These mRNAs generally conclude with a sequence of up to 200 adenosines, the **polyadenylic acid (polyA)** sequence at the 3' end. This sequence isn't coded by the DNA template for the gene; it is added **post-transcriptionally**. Not all mRNAs are polyadenylated. For example, histone mRNAs lack polyA tails. Polyadenylation seems to play a role in regulating the stability of mRNAs. An early event in the breakdown of some mRNAs is the removal of their polyA tails.

Transfer RNA

Transfer RNAs (tRNAs) have two functions that link the RNA and protein information systems. First, they must accept a specific amino acid, one of 20. They do this with accuracy greater than 99.99 percent, even distinguishing between chemically similar structures. But tRNAs share functions as well. The translating ribosome must be able to insert any of the 20 amino acids at the correct position in the growing polypeptide chain, with roughly the same efficiency. Otherwise, the number of proteins that a cell could make would be severely limited. This means that all tRNAs must have common structural features that are recognized by the ribosome.

You can see the common structural features of tRNAs at both the secondary and tertiary levels. Only a few sequences or bases are common to all tRNAs. The common secondary structure of tRNAs is the **cloverleaf** pattern, where the 5' and 3' sequences are base-paired, and then the other three stem-loops of the cloverleaf are formed by intramolecular base pairs over a short distance. All tRNAs end in the **acceptor sequence**, CCA, which furnishes a common structure for the ribosome to recognize. The CCA sequence is made either by transcription of the DNA template or added post-transcriptionally by an enzyme. The A of the acceptor CCA sequence does not have a 2' or 3' phosphate. Depending on the exact tRNA species, the amino acid is loaded onto either the 2' or 3' hydroxyl group of the acceptor A in an ester linkage. The other part of the tRNA is the **anticodon**, which forms base pairs with the trinucleotide codon sequence of the ribosome-bound message. The anticodon is found at the same place in each tRNA cloverleaf, away from the acceptor stem. This dual nature of tRNA provides a clue to its translation function: The acceptor stem accepts a specific amino acid, while the anticodon determines the placement of that amino acid at the correct point in the growing polypeptide chain.

The tertiary structure of all tRNAs are likewise similar. All known tRNAs are roughly L-shaped, with the anticodon on one end of the L and the acceptor stem on the other. Each stem of the L is made up of two of the stems of the cloverleaf, arranged so that the base pairs of each stem are *stacked* on top of each other. The parts of the molecule that are not base-paired are involved in other types of interactions, termed *tertiary* interactions. The tertiary structures of tRNAs thus reflect the dual functions of the molecule: The anticodons are well-separated from the acceptor stems. This feature allows two tRNA molecules to interact with two codons that are adjacent on an mRNA molecule. See Figure 10-4.

Figure 10-4

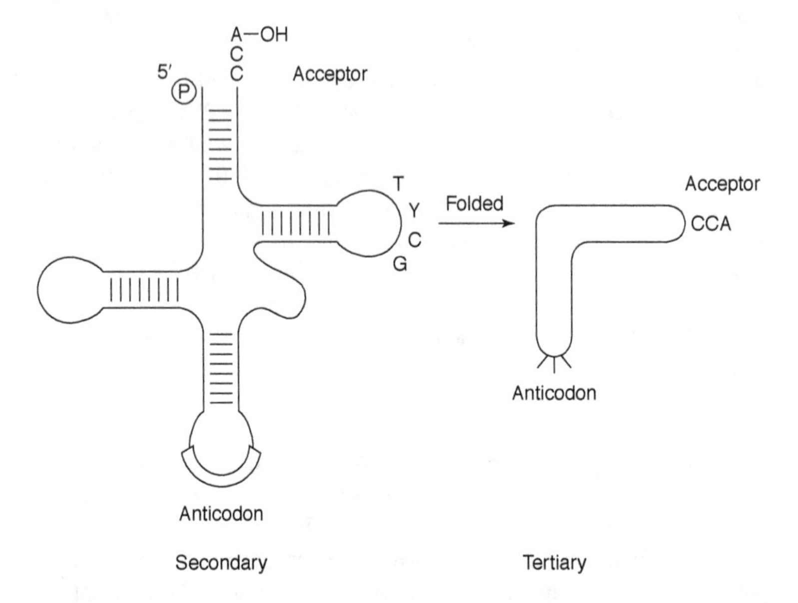

Ribosomal RNA

Ribosomal RNA is essential for protein synthesis. In fact, RNA is thought to be the catalytically active part of the very large complex of proteins and RNAs that synthesize proteins. Ribosomes and ribosomal RNAs are heterogeneous, with different sized rRNAs found in the small and large subunits of the ribosome. Ribosomes can be separated into two subunits. Each subunit contains both protein and RNA. Although they vary widely in size, ribosomal RNAs have common secondary structures. The larger size of the eukaryotic RNAs is due to their having extra structural domains inserted into the midst of the smaller ones, rather than by a totally new folding pattern.

Antibiotics are natural products, usually from soil bacteria and molds, which interfere with the growth of other bacteria. Often these antibiotics act on ribosomal RNA targets. For example, streptomycin, which has been used to treat tuberculosis, binds to a single region of bacterial 16S RNA, interfering with protein synthesis. The drug doesn't disrupt protein synthesis in humans, which allows for streptomycin's relatively high **therapeutic index**—the ratio of harmful to helpful doses of the drug. Conversely, bacteria can become resistant to antibiotics by changes in their rRNA, either by a change in the nucleotide sequence of the ribosomal RNA or by methylation of the rRNA.

Other Information-Processing RNAs

Other stable RNAs in the cell are involved in protein and RNA **processing**. Processing means changes in a protein after the synthesis of the peptide bond, or in an RNA after the synthesis of phosphodiester bonds during transcription. Some of these RNAs are catalytic.

Transcription

Transcription is the synthesis of RNA from a DNA template. The primary transcript may then be modified or processed to the final product. Eventually, the RNA product is degraded to nucleotides. While all of these reactions are potential sites for control of gene expression, most control of gene expression is transcriptional. This is an example of a general principle of biochemical control: Pathways are controlled at the first committed step, and RNA synthesis is the first committed step of gene expression.

Different RNAs are transcribed with different efficiencies, so that structural (transfer and ribosomal) RNAs are transcribed very efficiently. Each transcription cycle leads to a single molecule. In contrast,

one mRNA is translated into many protein molecules. This means that transcription of mRNA need not be so rapid.

Transcription in Prokaryotes

The most detailed molecular information about the transcription cycle is available in bacterial systems. The synthesis of RNA is initiated at the promoter sequence by the enzyme **RNA polymerase**. A single RNA polymerase type is responsible for the synthesis of messenger, transfer, and ribosomal RNAs.

When isolated from bacteria, prokaryotic RNA polymerase has two forms: The **core enzyme** and the **holoenzyme**. The core enzyme is a tetramer whose composition is given as $\alpha_2\beta\beta'$ (two alpha subunits, one beta subunit, and one beta-prime subunit). Core RNA polymerase is capable of faithfully copying DNA into RNA but does not initiate at the correct site in a gene. That is, it does not recognize the promoter specifically. Correct promoter recognition is the function of the holoenzyme form of RNA polymerase.

Figure 10-5

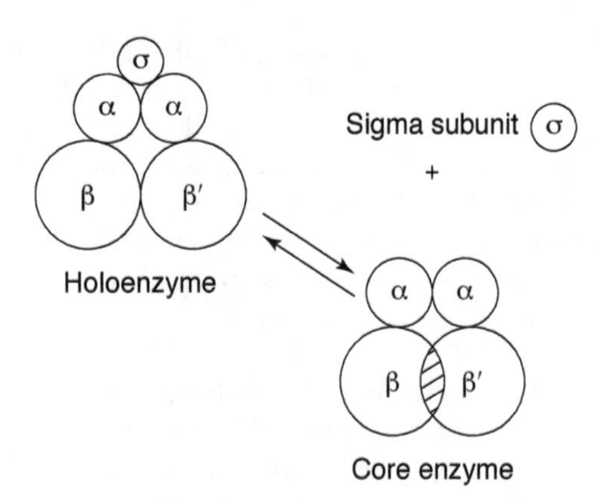

The RNA polymerase holoenzyme contains another subunit, σ (**sigma**), in addition to the subunits found in the core enzyme. Holoenzyme, $\alpha_2\beta\beta'\sigma$, is capable of correct initiation at the promoter region of a gene. Sigma thus must be involved in promoter recognition. Sigma subunits are related but distinct in different forms of RNA polymerase holoenzyme. These specialized σ subunits direct RNA polymerase to promoter sequences for different classes of genes. For example, bacteria exposed to high temperatures synthesize a set of protective proteins called **heat-shock proteins**. The genes for the heat-shock proteins have special promoter sequences that are recognized by an RNA polymerase holoenzyme with a specific σ subunit. The σ discussed here is the major σ of the common bacterium *E. coli*, about which most is known.

Promoter recognition

RNA polymerase holoenzyme starts by recognizing the promoter of a gene. The promoter isn't copied into RNA, but it is, nonetheless, an important piece of genetic information. The information in a promoter was determined by lining up a large number of promoters and counting how many times a particular base appeared at a given position in the various promoter sequences. The **consensus sequence** is given by the statistically most probable base at each point—the bases that appear most often in the promoter collection. Very few, if any, naturally occurring promoters match the consensus sequence exactly, but the "strength" of a promoter (how actively RNA polymerase initiates at it) correlates well with the degree of consensus match. For example, the promoters of genes for ribosomal RNA match the consensus well, while the promoters for the mRNA encoding some regulatory proteins match the consensus poorly. This correlates with the relative amounts of each gene product that are needed at any one time: many ribosomes, and only a few regulatory proteins.

The consensus sequence for an *E. coli* promoter has two conserved regions near positions -35 and -10 relative to the transcription start site. That is, the template-directed synthesis of RNA begins 35 base pairs downstream of the first consensus region and ten base pairs downstream of the second. The -35 consensus is:

TTG**ACA**.

The -10 consensus is:

TATAA**T**.

A couple of important points exist about the consensus. First, not all bases in the consensus are conserved to the same amount. The bases marked with bold type and underlined are more conserved than the others, and the -10 region is more conserved overall than is the -35 region. Secondly, the promoter sequence is asymmetrical; that is, it reads differently in one direction than in the other. (Compare this to the recognition sequence for the restriction enzyme BamHI, GGATCC.) This asymmetry means that RNA polymerase gets *directional information* from the promoter in addition to information about the starting point for transcription.

The transcription process

RNA polymerase only goes one direction from a promoter and only one strand of DNA is used as a template at any one time. To provide this template strand, the initiation of transcription involves a short unwinding of the DNA double helix. This is accomplished in a two-step fashion. First, RNA polymerase binds to the promoter to form the closed complex, which is relatively weak. Then, the double-stranded DNA goes through a conformational change to form the much stronger open complex through opening of the base pairs at the -10 sequence, as shown in Figure 10-6.

Figure 10-6

Transcription Initiation

1. Binding of RNA polymerase to template

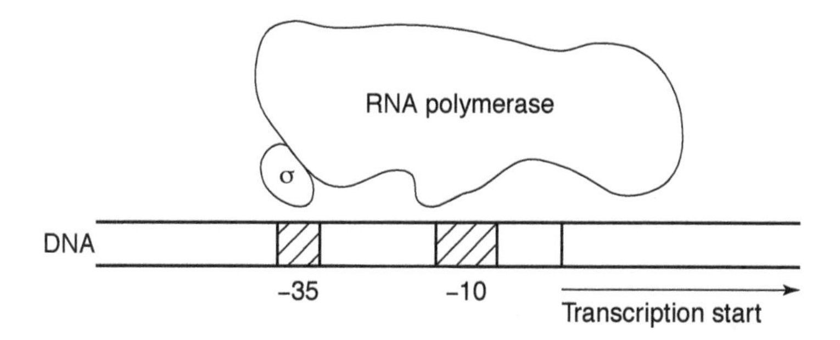

2. Dissociation of sigma from –35 and recognition of –10 sequence

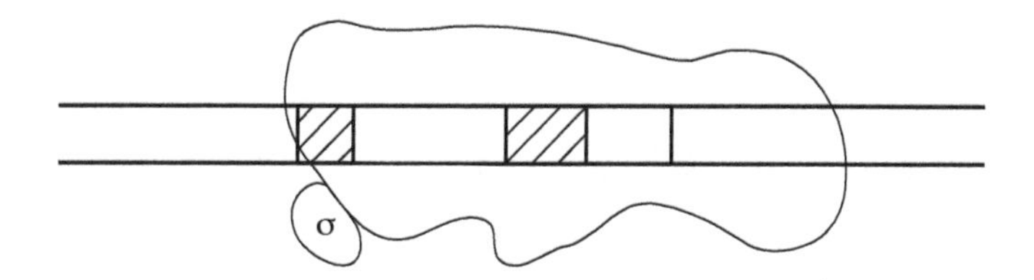

3. Establishment of open-promoter complex

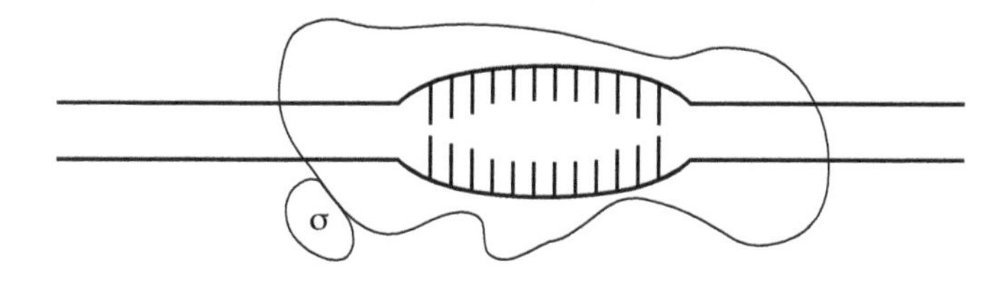

Consensus –10: TATAATG

Consensus –35: TTGACA

The initiator nucleotide binds to the complex and the first phosphodiester bonds are made, accompanied by release of σ. The remaining core polymerase is now in the elongation mode. Several experimental observations support the picture presented in the next figure, namely the fact that less than one σ exists in the cell per core enzyme in each cell.

Figure 10-7

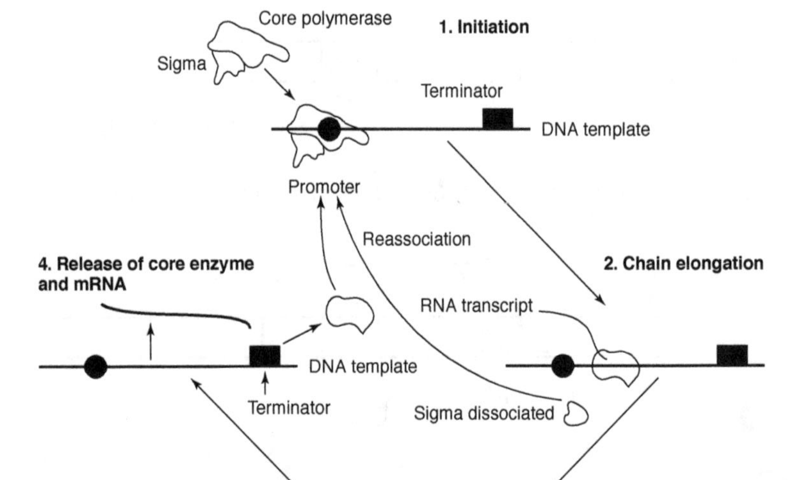

Elongation is the function of the RNA polymerase core enzyme. RNA polymerase moves along the template, locally "unzipping" the DNA double helix. This allows a transient base pairing between the incoming nucleotide and newly-synthesized RNA and the DNA template strand. As it is made, the RNA transcript forms secondary structure through intra-strand base pairing. The average speed of transcription is about 40 nucleotides per second, much slower than DNA polymerase. Other protein factors may bind to polymerase and alter the rate of transcription and some specific sequences are transcribed more slowly than others are. Eventually, RNA polymerase must come to the end of the region to be transcribed.

Termination of transcription *in vitro* is classified as to its dependence on the protein factor, rho (ρ). Rho-independent terminators have a characteristic structure, which features (a) A strong G-C rich stem and loop, (b) a sequence of 4-6 U residues in the RNA, which are transcribed from a corresponding stretch of As in the template. Rho-factor-dependent terminators are less well defined, as shown in Figure 10-8.

Figure 10-8

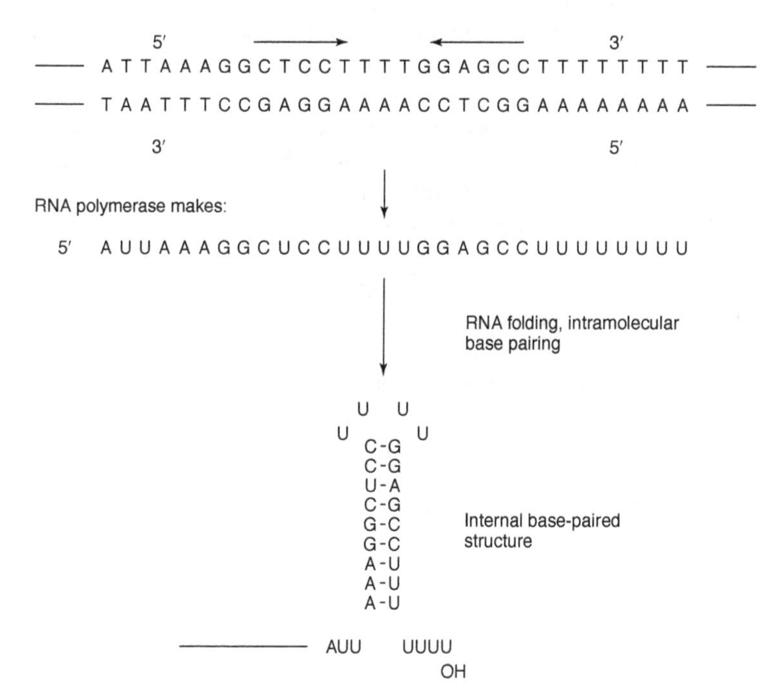

Rho factor: A protein that recognizes terminator regions.

Transcriptional Control in Bacteria

Because the first committed step of gene expression is the transcription of the gene, a large fraction of genetic control takes place at this step, especially through the initiation of transcription. Genetic control schemes may be classified in several ways.

The first method of classification is according to their physiological effect on gene expression, either positive or negative. In **positive control**, the control element or pathway stimulates gene expression. The effect of a mutation "knocking out" the control gene is therefore to decrease the level of gene expression. In this classification, a promoter sequence would be a positive control element, because removing it would decrease transcription. A positive control element is like the ignition switch in a car: removing it means that the car can't go. By contrast, in **negative control**, a control element acts to reduce or repress gene expression. The effect of a mutation "knocking out" a positive control gene would be to increase the level of expression. A negative control element is like the brakes in a car: removing it means that the car keeps moving. Several steps in the expression of a single gene may be controlled in positive and/or negative fashion. This can lead to finely tuned gene expression.

The location and manner in which a control element acts genetically determine a second classification. ***Cis*-acting** control elements affect gene expression from the same chromosome only. A promoter, for example, is a *cis*-acting element, because mutating it affects the level of gene expression from that chromosome only. In the preceding analogy, the brake on a car is *cis*-acting, because removing it affects the ability of only a single car to stop. ***Trans*-acting** control elements affect gene expression from a chromosome other than the one encoding them. For example, the σ subunit of RNA polymerase is *trans*-acting, because mutating it can affect the expression of genes from another chromosome. In general, but not always, *cis*-acting genes are DNA sequences that act as *sites* for regulation, and *trans*-acting genes code for proteins or other *diffusible* control molecules that act on these sites. By analogy, a red light is *trans*-acting because it stops several cars.

The third way in which gene expression schemes may be classified is by determining at which step of gene expression the control elements function. As mentioned previously, much of the overall genetic control in all organisms is transcriptional because RNA synthesis is the first committed step to gene expression. Control can also be post-transcriptional, that is, at any of the steps after transcription, including RNA processing, mRNA translation, protein metabolism, and so forth.

The Lactose Operon—A Case Study in Regulation

The biochemistry of the lactose (*lac*) operon explains many principles of regulation. The *lac* operon encodes a set of genes that are involved in the metabolism of a simple sugar, lactose. Lactose is a disaccharide composed of two sugars (galactose and glucose) with a β-linkage between carbon 1 of galactose and carbon 4 of glucose, as shown in Figure 10-9.

Figure 10-9

CH_2OH CH_2OH

O O O

Lactose

Operon Control

Negative control exerted by repressor

A. The repressor gene, *lacI*, is always active but at a low level.

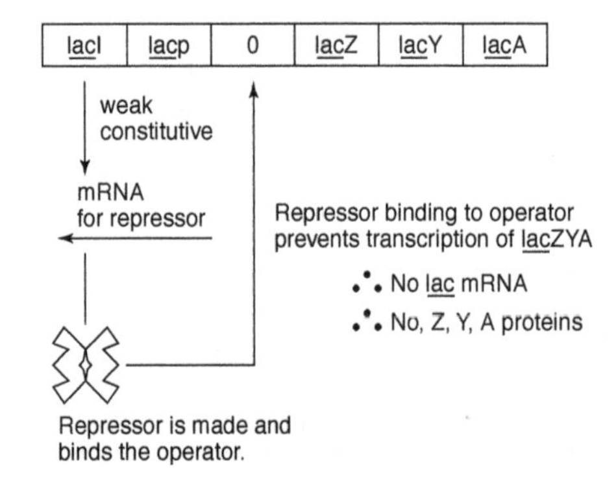

B. Repressor inactivated by binding with lactose metabolite

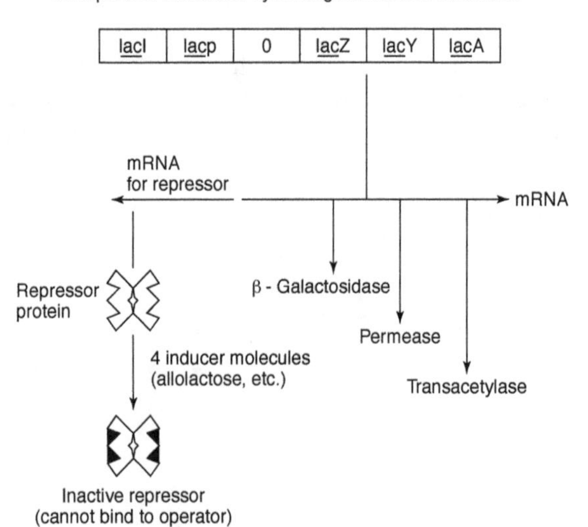

The *lac* operon encodes three proteins: β-galactosidase (the product of the *lacZ gene*), lactose permease (the product of the *lac*Y gene), and lactose transacetylase (the product of the *lac*A gene). The function of *lac*A is not known, but a mutation in either *lac*Z or *lac*Y means that the cell can't grow by using lactose as a sole carbon source. All three structural genes are transcribed from a common promoter site, in the direction Z-Y-A. The *lac* transcript is termed **polycistronic** because it contains more than one coding sequence.

Closely linked to the *lac* structural genes is the gene (*lac*I) for the *lac* **repressor**, a tetramer of four identical subunits. The repressor has two functions. First, it binds to the DNA near the *lac* promoter and prevents transcription of the structural genes. Secondly, it binds to a small molecule called an **inducer**. In the cell, the inducer is allolactose, a metabolite of lactose. The binding of inducer to the repressor is **cooperative**, meaning that the binding of one molecule of inducer makes binding of the next one more favorable, and so on. This means that the repressor binds inducer in an all-or-none fashion.

In the absence of inducer, the repressor protein binds to a sequence called the **operator** (*lac*O) which partially overlaps with the promoter. When it is bound to the operator, *lac* repressor allows RNA polymerase to bind the promoter and form an open complex but not to elongate transcription. Repressor is therefore a negative regulator of gene expression: If repressor is not present (for example, the bacterium is deleted for the *lac*I gene), then transcription of the *lac* genes occurs, and the structural genes are expressed whether or not inducer is present. The unregulated, "always on" expression caused by a *lac*I$^-$ mutation is called **constitutive expression**. This behavior is characteristic of a negative control element. *Lac*O is also a negative element. Deletion of *lac*O leads to constitutive expression of the *lac* genes. The difference between the two types of constitutive mutation is seen when the genes are put into an artificial situation where two copies of the relevant control genes exist. See Figure 10-10.

Figure 10-10

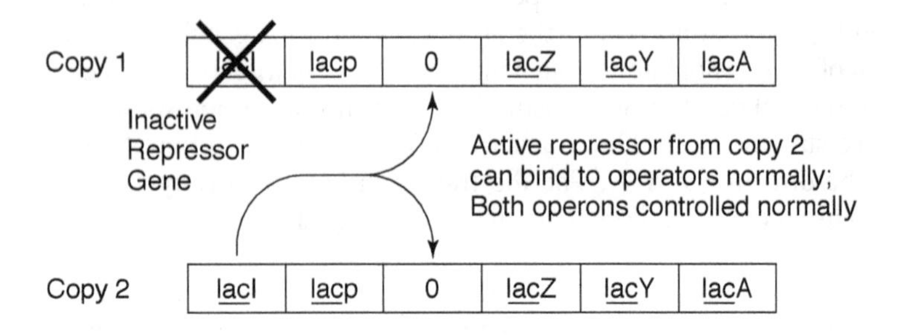

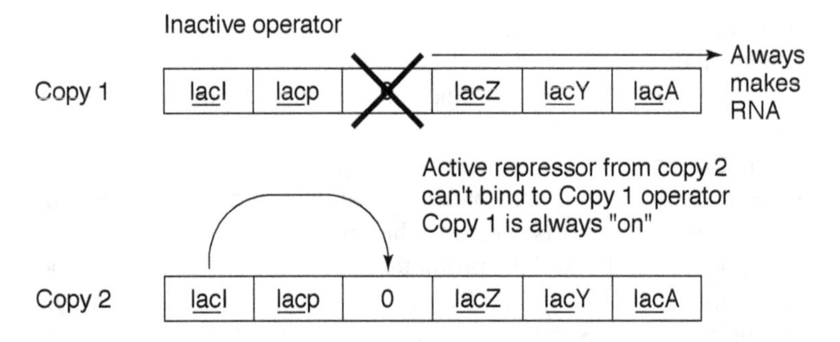

If a cell is provided with one wild-type and one mutant *lac*I gene, then the "good" (wild-type) copy of the repressor gene will provide functioning repressor to the other operator, and expression will be normally controlled. This behavior is characteristic of a **diffusible** control element, and the repressor is said to act in *trans*. If the same experiment is carried out with one wild-type and one mutant *lac*O gene, then the genes controlled by the mutant *lac*O will be constitutively expressed and the genes controlled by the wild-type copy of *lac*O will be normally regulated. This is characteristic of a site where the product of another gene acts, and *lac*O is therefore termed *cis*-acting.

This model also accounts for the behavior of mutants that cause repressor to fail to bind inducer. If the repressor can't bind inducer, then the *lac* genes it controls are permanently turned off because repressor will be bound to the operator whether inducer is present or not. In a diploid situation, both sets of genes will be turned off, because the repressor will bind to both operators.

A second level of control is superimposed on the repressor-operator interaction described previously. *In vivo, lac* gene expression is greatly reduced by the presence of glucose in the medium, even if enough lactose is present to release the repressor from the operator. This makes good metabolic sense. Glucose is more easily catabolized (broken down) than is lactose, which must be broken down into glucose and galactose, followed by the specialized pathway for galactose metabolism. The phenomenon by which glucose reduces the expression of the *lac* operon is called **catabolite repression**, as shown in Figure 10-11.

Figure 10-11

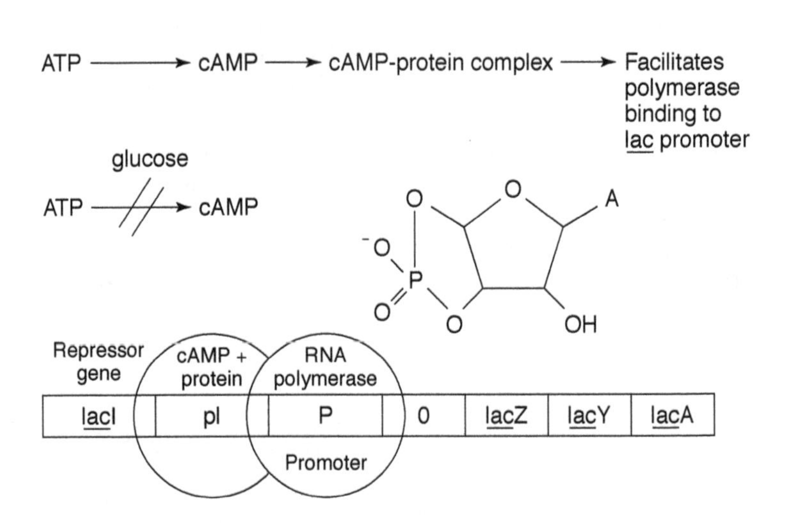

Catabolite repression is a two-part system. The first component is the small-molecule regulator, **cyclic AMP**. Glucose decreases cyclic AMP synthesis. The second component is cyclic AMP binding protein, **CAP**. CAP binds cAMP and thereby helps RNA polymerase bind to the promoter. When bound to cAMP, CAP binds to a sequence at the 5' end of the *lac* promoter. CAP binding bends the DNA, allowing protein-protein contact between CAP and polymerase. It therefore behaves in the opposite manner of repressor. Repressor (LacI) binds to operator DNA only in the absence of its small-molecule ligand, while CAP binds to promoter DNA in the presence of its small-molecule ligand.

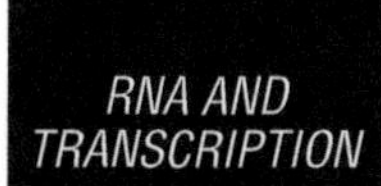

These two complementary systems allow the bacterial cell to metabolize lactose in response to *two* stimuli. "Switching on" the expression of the *lac* operon requires both the absence of glucose and the presence of lactose. This series of switches allows complex expression patterns to be built up from simple components. For this reason, the *lac* system is a model for other, apparently more complex, biological control systems, such as hormone action or embryonic development.

CHAPTER 11
PROTEIN SYNTHESIS

Transcription and Translation

The synthesis of RNA is called **transcription** because it is simply the copying of DNA "language" into RNA. Like the transcription of spoken language into written language, the units of information (nucleotides in nucleic acids, words in speech and writing) are the same. Translation—the conversion of one language to another—is much more difficult, whether in human language or in biochemical language. Translation can't be too literal and has to preserve the context of information as well as its symbols. The information problem of biological translation is the way in which a protein sequence can be encoded by a nucleic acid sequence.

The correspondence between nucleic acid information and protein information is given by the **genetic code**. which is a set of rules giving the correspondence between mRNA and protein sequence information. See Figure 11-1.

The genetic code can be thought of as a dictionary giving the equivalents for words from one language to another. However, the dictionary isn't enough. Just as the translation of one language into another requires a translator, the genetic code requires an adaptor molecule. Transfer RNA is that adaptor molecule for biological information. Secondly, the error frequency of the process must be kept to a minimum. The wrong amino acid in a protein could, in principle, lead to the death of the cell, just as the wrong word in translation of a diplomatic message could lead to a war. Both cases need a proofing mechanism to check that the information transfer is accurate. Thirdly, punctuation and reading frame selection are essential components of the process. Because the genetic code is a triplet code, two of the three "messages" in a string of nucleic acids are usually meaningless. In fact, a number of severe genetic diseases are caused by mutations that cause a "frameshifted" protein whose information is meaningless. A biological translation system must know where messages start and stop.

Figure 11-1

The Genetic Code

First letter (rows); Second letter (columns)

First letter	U	C	A	G
U	UUU, UUC } Phe UUA, UUG } Phe	UCU, UCC, UCA, UCG } Ser	UAU, UAC } Tyr UAA, UAG } STOP	UGU, UGC } Cys UGA } STOP UGG } Trp
C	CUU, CUC, CUA, CUG } Leu	CCU, CCC, CCA, CCG } Pro	CAU, CAC } His CAA, CAG } Gin	CGU, CGC, CGA, CGG } Arg
A	AUU, AUC, AUA } Ile AUG } Met	ACU, ACC, ACA, ACG } Thr	AAU, AAC } Asn AAA, AAG } Lys	AGU, AGC } Ser AGA, AGG } Arg
G	GUU, GUC, GUA, GUG } Val	GCU, GCC, GCA, GCG } Ala	GAU, GAC } Asp GAA, GAG } Glu	GGU, GGC, GGA, GGG } Gly

A protein molecule's amino acid sequence determines its properties. This has been shown for many proteins, which can be denatured and then refolded to re-form active protein tertiary structure. The biological translator thus has a somewhat easier task than the translation of human languages, because the mRNA and protein sequences are colinear. Important parts of the information don't rearrange from one language to another in contrast to the way, for example, that verbs occur at different positions in German and English sentences.

The conversion of nucleic acid into protein information doesn't completely solve the problem of translation. Proteins must be targeted to their appropriate locations, either inside or outside the cell. In eukaryotes especially, proteins must be broken down at appropriate rates; some proteins have longer half-lives than others do. These steps are all possible points for cellular control.

Fidelity in tRNA Aminoacylation

Aminoacylation is a two-step process, catalyzed by a set of enzymes known as **aminoacyl-tRNA synthetases**. Twenty aminoacyl-tRNA synthetases reside in each cell, one per amino acid in the genetic code. In the first step of aminoacyl-tRNA synthesis, ATP and the appropriate amino acid form an *aminoacyl adenylate* intermediate. Inorganic pyrophosphate is released and subsequently broken down to free phosphate by the enzyme inorganic pyrophosphatase. The aminoacyl adenylate is a "high-energy" intermediate, and in the second step, the transfer of amino acids to the acceptor end of tRNA occurs without any further input of ATP, as shown in Figure 11-2.

Figure 11-2

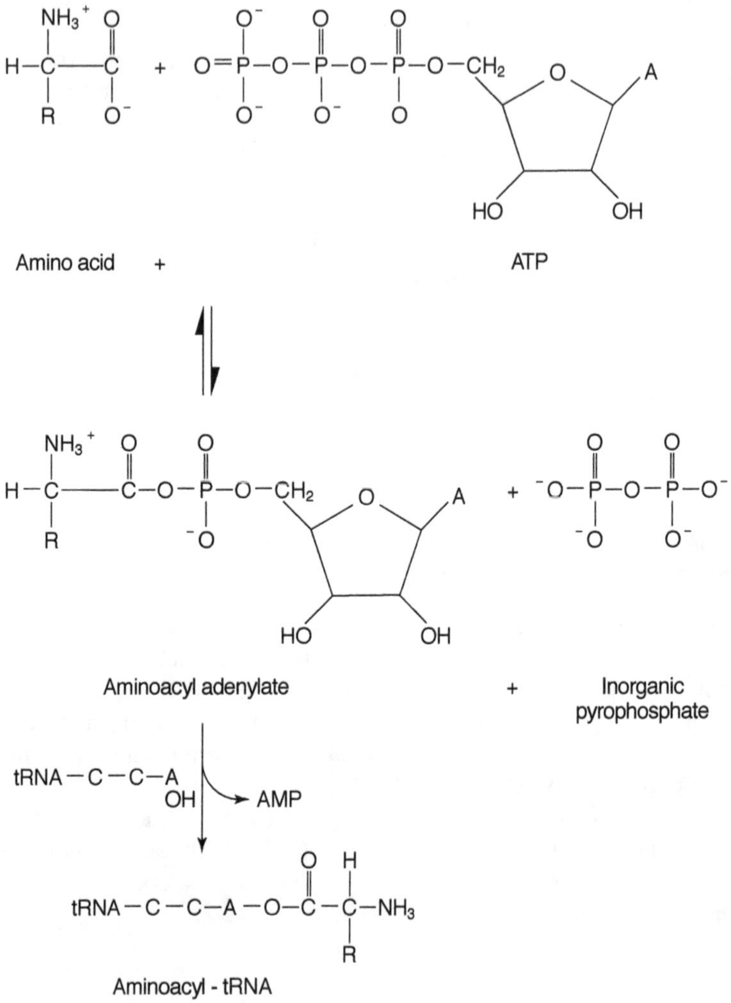

Amino acid + ATP

$\rightleftharpoons$

Aminoacyl adenylate + Inorganic pyrophosphate

tRNA—C—C—A OH → AMP

Aminoacyl - tRNA

The aminoacyl-tRNA synthetase carries out an **editing** step to ensure against misacylated tRNA being used in protein synthesis. Because the ribosome must treat all aminoacyl-tRNAs as the same for making the peptide bond, any tRNA bearing the wrong amino acid would be used for protein synthesis, possibly causing the synthesis of a harmful protein. Editing aminoacyl-tRNAs for accuracy is carried out by a second active site on the aminoacyl-tRNA synthetase. (See Figure 11-3.)

Figure 11-3

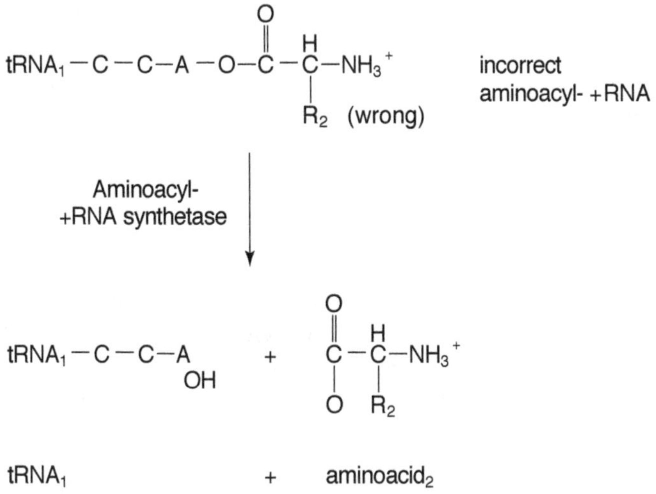

$$tRNA_1-C-C-A-O-\overset{\overset{\displaystyle O}{\|}}{C}-\underset{\underset{\displaystyle R_2}{|}}{\overset{H}{C}}-NH_3^+ \quad \text{(wrong)}$$

incorrect aminoacyl-tRNA

Aminoacyl-tRNA synthetase ↓

$$tRNA_1-C-C-A\underset{OH}{} \quad + \quad \underset{\underset{\displaystyle O^-}{|}}{\overset{\overset{\displaystyle O}{\|}}{C}}-\underset{\underset{\displaystyle R_2}{|}}{\overset{H}{C}}-NH_3^+$$

$tRNA_1$ + $aminoacid_2$

This function of the enzyme cleaves the incorrect aminoacyl-tRNA to yield free amino acid and tRNA. This process is analogous to the editing by the 3'-5' exonucleolytic function of DNA polymerases that goes on during DNA synthesis. Like that process, aminoacyl-tRNA editing causes a "futile cycle" in which the enzyme synthesizes a bond by using energy and then breaks it down. In both cases, the fidelity of information processing is preserved at the expense of energy "wastage." Mistakes are so dangerous to the cell that the expenditure is a good bargain.

Initiation of Protein Synthesis

The protein synthetic machinery must select the appropriate starting points for mRNA reading and peptide bond formation. AUG is usually used as the starting codon, and essentially all proteins begin with a methionine. AUG is also the codon for methionine that occurs in the interior of a protein as well, so there must be a mechanism to distinguish between the two types of methionine codons.

The steps of initiation occur on the isolated small subunit (30S) of the prokaryotic ribosome. Ribosomes contain two subunits, a 30S and 50S subunit, which associate to form a 70S particle. (The S values refer to the rate at which each component sediments in the ultracentrifuge; they don't always add up.) In general, the 30S subunit is mostly involved in the decoding and tRNA-mRNA interaction process, while the 50S subunit is involved in actual peptide bond synthesis. Ribosomal subunits are dissociated prior to the initiation reaction.

Translation is initiated at the 5' end of mRNA. Because RNA is synthesized in a 5'-3' direction, a bacterial mRNA can start translation while the 3' sequences are still being transcribed. This is important in several forms of biological control.

A special initiator tRNA, **$tRNA^{met}_I$** (I stands for initiator) is used for beginning protein synthesis. In bacteria, this initiator tRNA carries the modified amino acid N-formylmethionine (fmet). The formylation reaction transfers the formyl group from formyl-tetrahydrofolate to methionyl-$tRNA^{met}{}_{I}{}^{+}$. This initiator tRNA is used to recognize initiation codons; it does not insert met in response to an internal AUG codon. As a further safeguard, the formylation reaction ensures that the initiator methionine can only be at the amino terminus of the synthesized protein.

The decoding step of protein synthesis involves **base-pairing** between mRNA codon and tRNA anticodon sequences. A further base-pairing event between noncoding regions of mRNA and rRNA is required to select the proper reading frame and initiation codon. Bacterial mRNAs contain a purine-rich sequence (called a 'Shine-Dalgarno' or **RBS**, which is an abbreviation of Ribosome-Binding Sequence) in the 5' nontranslated region of the mRNA. This sequence is complementary to the 3' end of the small subunit rRNA, 16S rRNA. See Figure 11-4.

Figure 11-4

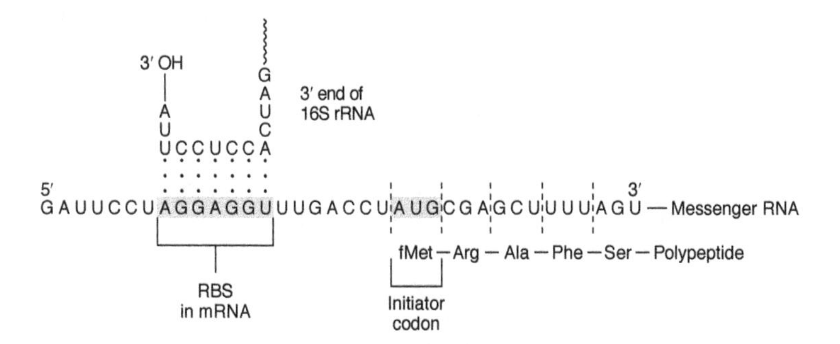

After base-pairing is established, protein synthesis starts with the first AUG downstream of the RBS. This feature of initiation is used as a form of translational control. Messenger RNAs with the greatest degree of RBS complementarity to 16S rRNA are translated most efficiently, presumably because they initiate more efficiently.

Several protein **factors** are involved in the initiation process. These factors aren't usually part of the ribosome; instead, they help form an active initiation complex. Initiation factor 3 (IF3) helps keep the 30S subunit dissociated from the 50S subunit and available for protein synthesis. IF1 binds to the isolated 30S subunit and helps form the complex between the RBS and 16S rRNA. IF2 forms a complex with fmet-$tRNA^{met}_{I}$ and GTP, releasing IF3. After the complex contains mRNA and initiator fmet-tRNA, the following things occur: GTP is hydrolyzed to GDP, the initiation factors are released from the ribosome, and the 50S subunit associates with the complex to form an elongating ribosome, as shown in Figure 11-5.

Elongation

The elongating ribosome essentially carries out the same step, peptide bond synthesis, over and over until a termination codon is reached. Several sites exist for binding aminoacyl-tRNAs on each ribosome. For the purposes of this book, however, only two are of importance. The **A (acceptor) site** is where the incoming aminoacyl-tRNA is bound to the ribosome. The initial step of decoding by codon-anticodon base-pairing occurs here. The **P (peptidyl) site** is where the peptidyl- (or initiator fmet-) tRNA is bound. Peptide bond synthesis involves the transfer of the peptide or fmet from the P site tRNA onto the free amino group of the A site aminoacyl-tRNA. Note how this means that the protein is synthesized in the amino- to the carboxyl- direction (N-C). After the peptide bond is formed, the ribosome **translocates**, moving the new peptidyl-tRNA from the A site to the P site, and the cycle begins again. See Figure 11-6.

Figure 11-5

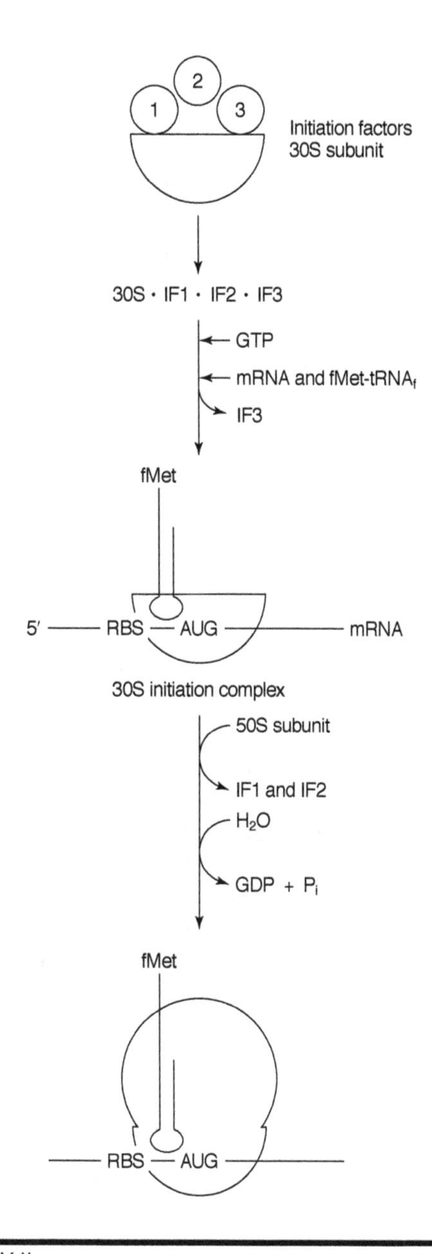
2
1
3
Initiation factors
30S subunit
30S · IF1 · IF2 · IF3
GTP
mRNA and fMet-tRNA$_f$
IF3
fMet
5′
RBS
AUG
mRNA
30S initiation complex
50S subunit
IF1 and IF2
H_2O
GDP + P_i
fMet
RBS
AUG

Figure 11-6

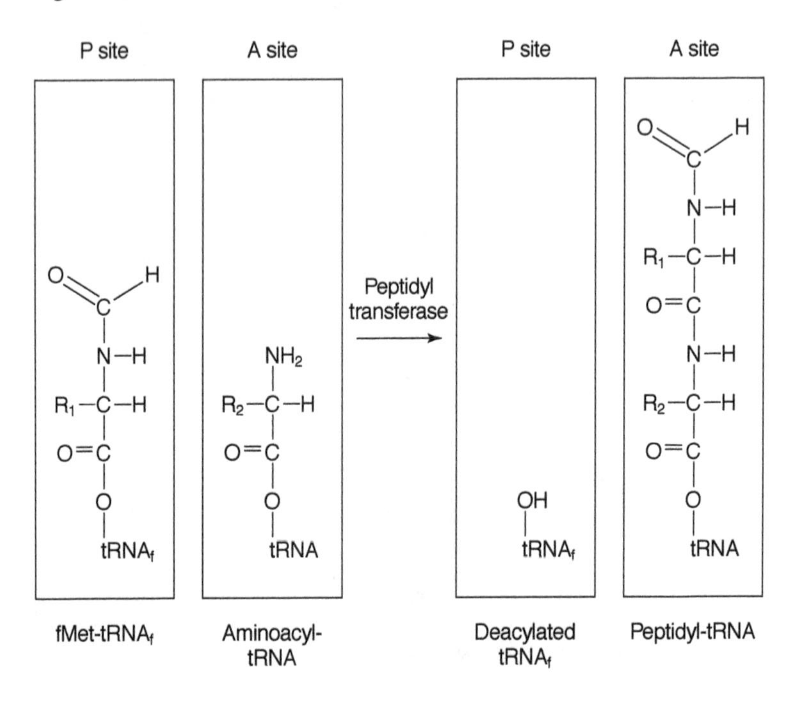

The elongation process is carried out with the assistance of **elongation factors** that use GTP to deliver the new aminoacyl-tRNA to the ribosomal A site. EF-Tu (u stands for "unstable") binds to aminoacyl-tRNA and GTP. After it is bound to the ribosome, EF-Tu hydrolyzes the GTP to GDP and inorganic phosphate. A second factor, EF-Ts (s stands for "stable") binds to the complex of GDP and Ts, causing GDP to be released from the factor so it can be replaced by GTP. Ts is therefore a guanine nucleotide **exchange factor**; it is displaced from Tu when GTP binds. Tu is now available for aminoacyl-tRNA binding, as shown in Figure 11-7.

Figure 11-7

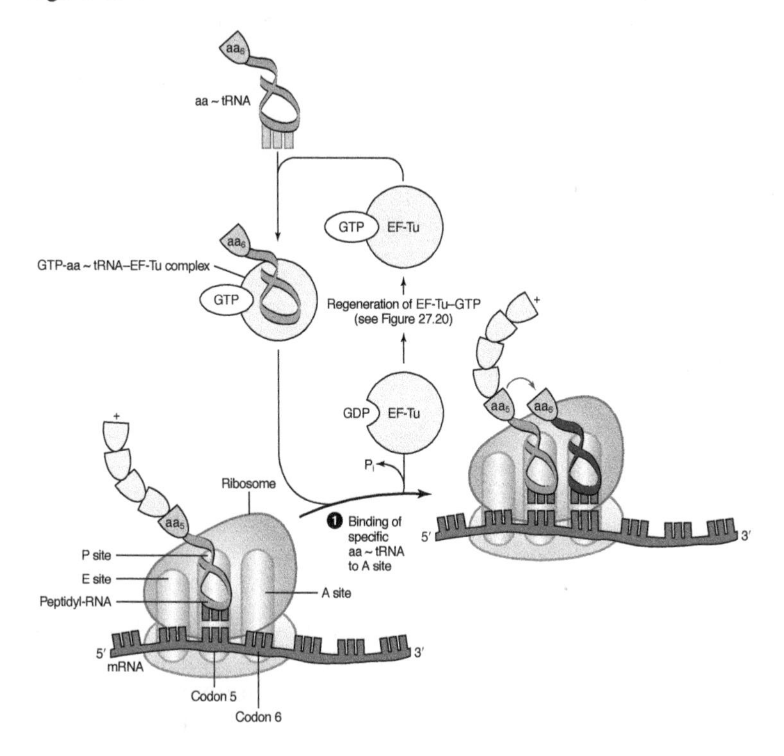

The 23S rRNA of the large ribosomal subunit catalyze the actual process of **peptide bond formation**. The synthesis of the peptide bond requires no energy input; it occurs because the aminoacyl bond of the tRNA at the P site is itself a "high energy" bond, with a free energy of hydrolysis essentially equal to that of an ATP phosphate. The peptide bond is more stable to hydrolysis, meaning that energy flows "downhill" during the process just as in the formation of amides from active esters in other organic reactions.

Translocation, however, requires the input of energy (again, in the form of GTP) with the participation of the elongation factor EF-G. The translocation reaction moves the peptidyl-tRNA from the A-site to the P-site. The uncharged tRNA is removed from the P-site (it remains bound at an Exit or E-site for another cycle of elongation), while the ribosome and mRNA move relative to each other. This is shown in Figure 11-8.

Figure 11-8

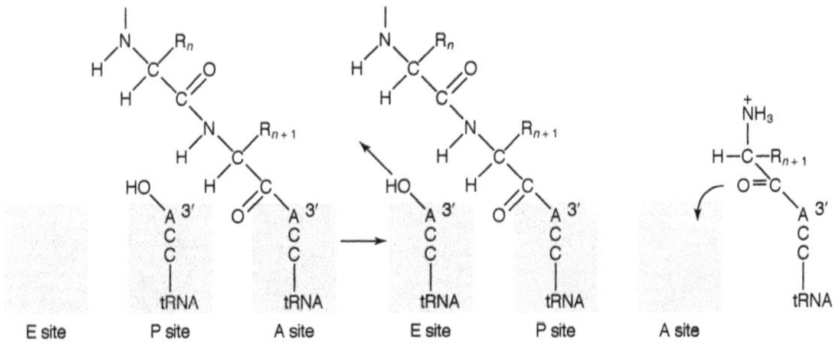

The A-site is free to accept the next aminoacyl-tRNA bound to elongation factor T_u. The growing peptide chain folds while still on the ribosome. As the ribosome moves down the mRNA chain, the initiation region (RBS) of the mRNA becomes available for reinitiation. This leads to the formation of a single mRNA with many ribosomes bound to it, called a **polysome**, as shown in Figure 11-9.

Termination of Translation

Three of the 64 codons, UAG, UAA, and UGA, do not specify any amino acid. When a translating ribosome encounters such a stop codon, no amino acid is inserted. Instead, one of two **release factors** binds to the stalled ribosome and causes the release of peptidyl-tRNA. The release factors are codon-specific. RF1 causes termination at UAA and UAG, while RF2 recognizes UAA and UGA. A third factor, RF3, is a GTPase and helps dissociate the complex. The ribosomal subunits are now separated from each other in preparation for the next initiation event. The binding of initiation factors to the 30S subunit helps this.

Protein synthesis is an energy-intensive process. High-energy phosphate bonds are expended for each peptide bond formed. One high-energy bond is consumed when an amino acid is activated by its aminoacyl-tRNA synthetase. Delivery of aminoacyl-tRNA by EF-Tu consumes one GTP per amino acid, and the translocation reaction consumes another.

Figure 11-9

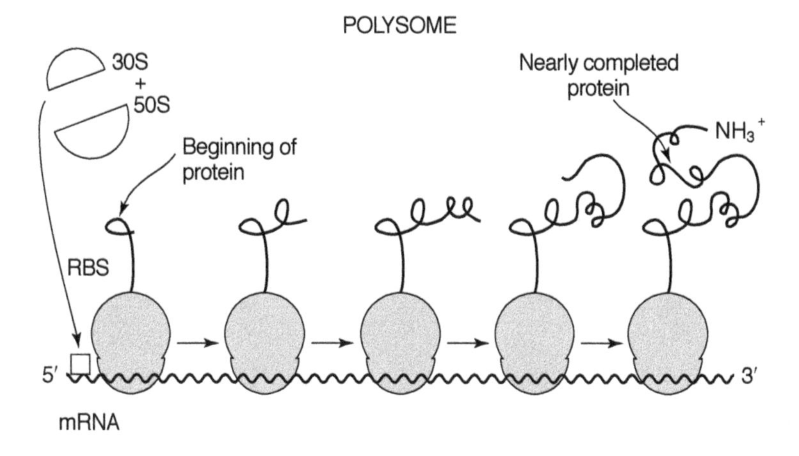

CHAPTER 12
EUKARYOTIC GENES

Eukaryotic Information Flow

The previous chapters concerned the flow of genetic information in prokaryotes—especially bacteria such as *E. coli*. Some of these processes are tightly coupled so that, for example, ribosomes can begin translating an mRNA before it is completely transcribed. In contrast, organisms with a nucleus must, by definition, have their informational molecules in separate compartments. In terms of the Central Dogma (Chapter 4, Volume 1), DNA is in the nucleus; its RNA must be in both the nucleus and the cytoplasm, and proteins must be made in the cytoplasm. After they are made, however, they can function in any part of the cell or be secreted. The fact that many organisms contain over a thousand times more DNA than do bacteria introduces a further complication. This information must be copied, localized, and expressed in a coordinated and regulated fashion.

DNA and Genomes in Eukaryotes

The total amount of DNA in an organism (its **genome**) can be estimated by physical measurements. Three basic classes of DNA exist in higher organisms. The term "complexity" refers to the number of independent sequences in DNA. Eukaryotic DNA can be divided into several classes of complexity. About half of the total DNA in a mammal is found in the most complex fraction. This fraction of the genome codes for functional genes and corresponds to sequences that exist in only one copy per genome. About a fifth of the DNA is moderately repetitive and present on the order of hundreds to thousands of times per genome. This fraction includes some sequences that are transcribed from many copies of the same sequence. For example, the genes for ribosomal RNA reside in this fraction. The remainder of the DNA is highly repetitive and can occur on the order of millions of copies per genome. This DNA is not transcribed much at all and may include DNA that is involved in chromosome structure.

Each of the three fractions contain a number of sequences that are sometimes called "junk" and can represent, for example, viruses that found their way into DNA in the past but were inactivated, leading to the fact that these sequences remain in the genome, but never express themselves. All of this DNA must be highly condensed. The DNA in each chromosome is a single molecule, on the order of several centimeters in length; the total DNA in a eukaryotic cell is as much as three meters long. This DNA must be condensed so as to fit into a nucleus that is about 10^{-5} meters (10vm) in diameter. The condensed structure of eukaryotic DNA is called **chromatin**.

Structure of Chromatin

Several levels of chromatin organization exist. Cytologically, chromatin classes may be distinguished by appearance when stained for light microscopy. **Heterochromatin** remains condensed throughout the cell cycle; microscopically, it looks "clumped" in the nucleus. Heterochromatin sequences are not transcribed. **Euchromatin** refers to the chromatin that appears less condensed in the microscope. Transcribed genes are found in euchromatin. Two types of heterochromatin exist. **Facultative heterochromatin** is present in the nucleus of some cells, but not in all. For example, in animals, the genes encoding β-globin are condensed in cells that are not precursors to blood cells. The term **constitutive heterochromatin** refers to DNA sequences that are condensed in all cells of an organism. Constitutive heterochromatin is associated with highly reiterated DNA. The DNA in centromeres and telomeres (the parts of the chromosome that attach to the mitotic spindle and the ends, respectively) is found in constitutive heterochromatin.

Chromatin structure is organized at several levels. The basic structure of chromatin—either heterochromatin or euchromatin—is called the **nucleosome**. The nucleosome is a complex of 146 base pairs of DNA, wound in two turns around the outside of a disk-like complex of eight proteins (called **histones**). The histone core contains two copies each of four histones, H2A, H2B, H3, and H4. The histone octamer is wrapped by very close to two turns of DNA. Linker DNA and another histone (H1) join together the nucleosomes (about 65 base pairs' worth). H1 binds cooperatively to nucleosomes, so that a gene can be "zipped up" all at once by the binding of many H1 molecules successively. See Figure 12-1.

Figure 12-1

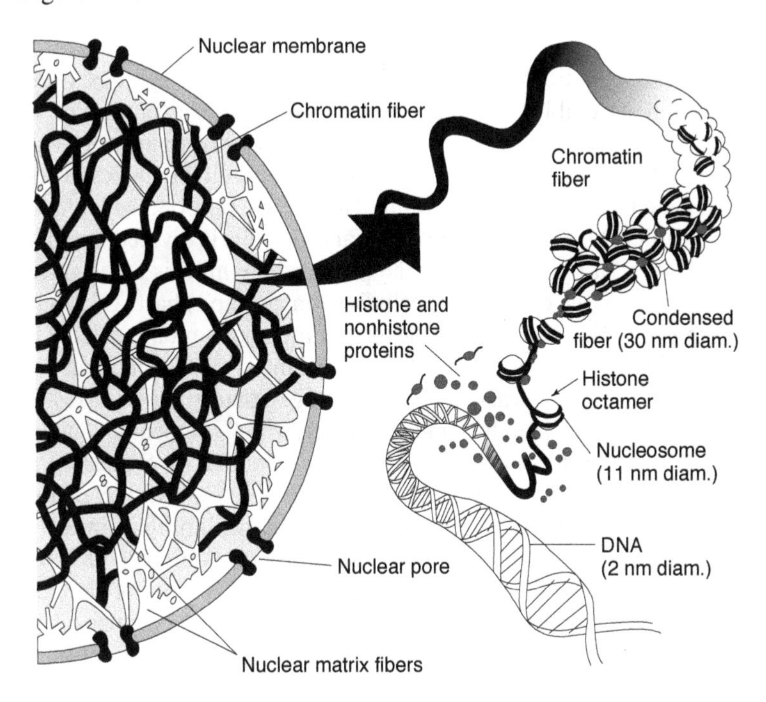

Nucleosomes are packed into more compact structures. First, the nucleosome fiber is condensed in a helix-like array, called the 30 nm fiber (nm refers to a nanometer, which is 10^{-9} meters). Scaffolding proteins hold the chromatin domains together. For active genes, these domains form loops, with the active sequences on the brushy part of the loops, and the regions between genes attached to the scaffold. Nonhistone proteins regulate the condensation and decondensation of chromatin. Again, active genes (those that are being transcribed) are present in the less-condensed portions of the chromatin.

Chromatin Replication

Eukaryotic genetic replication involves both DNA synthesis and chromatin assembly. Chromosomal DNA synthesis is similar to prokaryotic DNA replication in that each of the two strands serves as template for new synthesis. In contrast to the situation in prokaryotes, eukaryotic DNA replication is limited to a single portion, the S phase, of the **cell cycle**. The cell cycle of eukaryotic cells is divided into distinct phases. M phase, which is the phase where mitosis takes place, is the "start" of the cycle. After cell division, a "gap" phase, G1, commences, in which enzyme synthesis and metabolism take place. G1 can last for a very long period of time, and many cells in animals are "arrested" in G1 for years without dividing. Controlling G1 arrest is clearly important for understanding cancer, which is essentially a disease of replication control. The movement from G1 to S phase (the DNA replication phase) commits the cell to dividing. After S phase, a second gap exists, G2, which lasts until the beginning of mitosis. The biochemical reactions that govern these events involve proteins that are made and then broken down at specific points of the cell cycle. These proteins, called **cyclins**, are kinases that phosphorylate other proteins in the cell, leading eventually to the start of chromosomal DNA replication. At other phases of the cell cycle, other cyclin-type kinases control the entry into mitosis and the various phases of mitosis itself. See Figure 12-2.

Figure 12-2

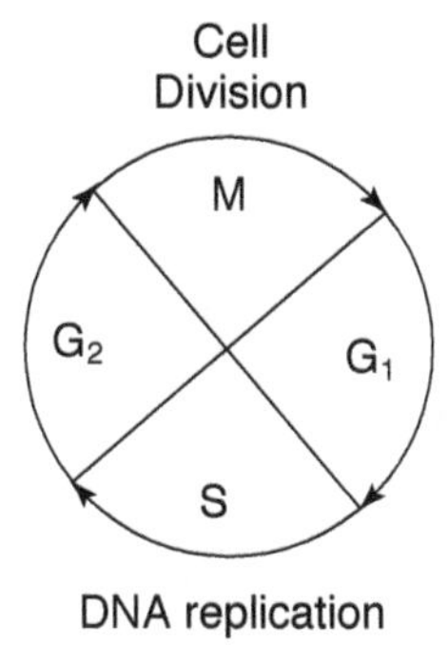

Unlike prokaryotes, where DNA replication begins at a single origin, eukaryotic cells use **multiple origins** of replication to initiate bidirectional synthesis. The replication forks appear to be attached to the nuclear membrane at distinct sites; the replicating chromatin may be pulled through these sites. A number of DNA polymerases exist in the nucleus. Activation of replication involves the assembly of one form of the enzyme, polymerase δ, with a special subunit called proliferating cell nuclear antigen (PCNA) so that it is capable of synthesizing long chains of DNA without falling off the template strand. As in prokaryotic DNA replication, continuous and discontinuous replication occurs on the leading and lagging strands respectively, so that the overall process is semiconservative. See Figure 12-3.

Nucleosome assembly differs from DNA replication. Histone content of a cell doubles during cell doubling, just as the DNA content does. On the leading strand, the pre-existing histone octamers briefly dissociate from the template and then re-bind to the double helix. Newly made nucleosome core particles associate with DNA on either strand. Thus, the overall process of histone doubling is conservative (the histones stay together during replication), in contrast to DNA synthesis, which is semiconservative.

Figure 12-3

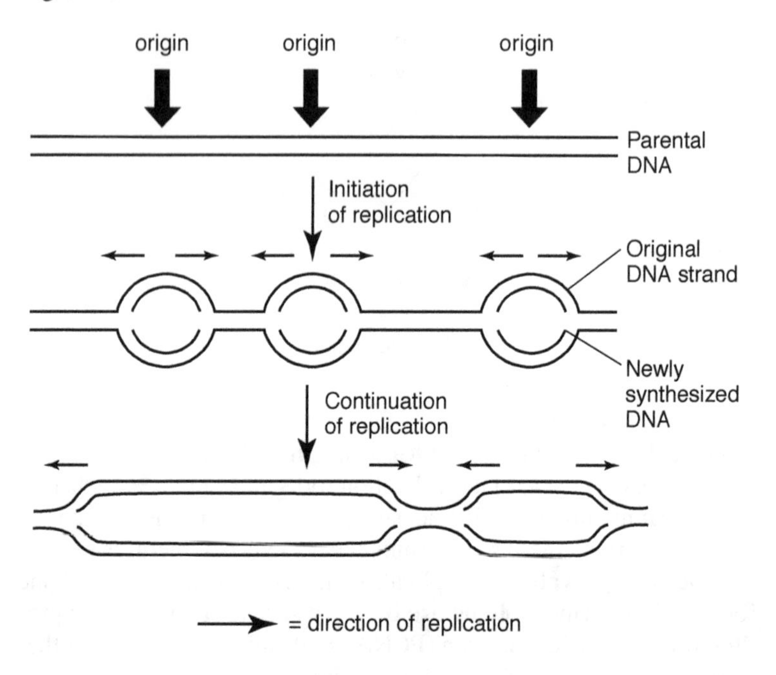

Figure 12-4

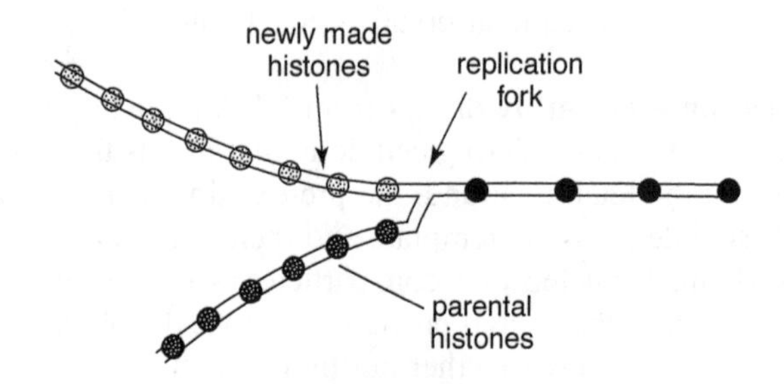

The end of a linear chromosome is called a **telomere**. Telomeres require a special mechanism, because the ends of a linear chromosome can't be replicated by the standard DNA polymerases. Replication requires both a template and a primer at whose 3' end synthesis begins. The primer can't be copied by the polymerase it primes. What copies the DNA complementary to the primer? In a circular chromosome, the primer site is to the 3' direction of another polymerase, but in a linear chromosome, no place exists for that polymerase to bind. As a result, unless a special mechanism for copying the ends of chromosomes is used, there will be a progressive loss of information from the end of the linear chromosome. Two characteristics about telomeres help avoid this situation. First, they consist of a short sequence—for example, AGGGTT—repeated many times at the end of each chromosome. Telomeres, therefore, are part of the highly repetitive DNA complement of a eukaryotic cell. Secondly, a specific enzyme, telomerase, carries out the synthesis of this reiterated DNA. Telomerase contains a small RNA subunit that provides the template for the sequence of the telomeric DNA. Eukaryotic somatic cells have a lifespan of only about 50 doublings, unless they are cancerous. One theory holds that a lack of telomerase in cells outside the germ line causes this limitation.

Reverse Transcription

Reverse transcription (which occurs in both prokaryotes and eukaryotes) is the synthesis of DNA from an RNA template. A class of RNA viruses, called **retroviruses**, are characterized by the presence of an RNA-dependent DNA polymerase (reverse transcriptase). The virus that causes AIDS, Human Immunodeficiency Virus (HIV), is a retrovirus. Because nuclear cell division doesn't use reverse transcriptase, the most effective anti-HIV drugs target reverse transcriptase, either its synthesis or its activity. Telomerase, discussed in the previous section, is a specialized reverse transcriptase enzyme. See Figure 12-5.

Figure 12-5

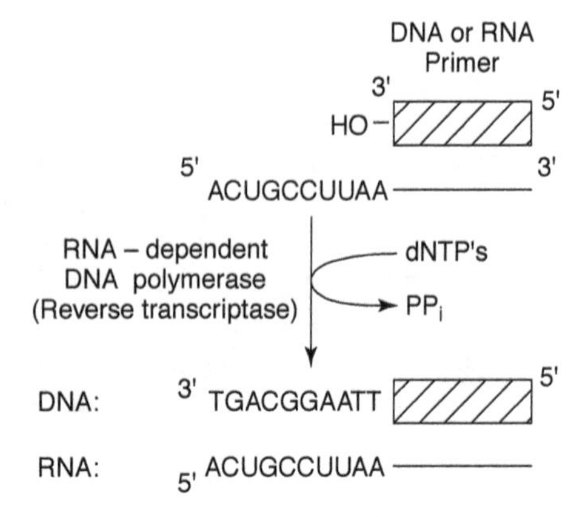

Like other DNA polymerases, reverse transcriptases are primer- and template-dependent. They also possess an RNase H activity (H stands for hybrid) that can degrade the RNA template after it is used for synthesis of the first DNA strand. The enzyme then can copy the first strand of DNA to make a double-stranded molecule.

Reverse transcription is error prone relative to DNA replication because reverse transcriptases don't have an editing (3'-5') exonucleolytic activity. This has one important consequence for HIV treatment and epidemiology. HIV mutates very rapidly. In advanced AIDS patients, the virus that is isolated from the bloodstream often bears very little resemblance to the original infecting strain. This rapid variation means that drug-resistant mutant strains of the virus arise frequently, and drug treatment doesn't work well. Secondly, the rapid mutation rate complicates vaccine development—new strains that are not neutralized by the vaccine can appear and infect individuals that were vaccinated against the original strain.

Eukaryotic Gene Structure

Although humans contain a thousand times more DNA than do bacteria, the best estimates are that humans have only about 20 times more genes than do the bacteria. This means that the vast majority of eukaryotic DNA is apparently nonfunctional. This seems like a contradiction. Why wouldn't more complicated organisms have more DNA? However, the DNA content of an organism doesn't correlate well with the complexity of an organism—the most DNA per cell occurs in a fly species. Other arguments suggest that a maximal number of genes in an organism may exist because too many genes means too many opportunities for mutations. Current estimates say that humans have about 100,000 separate mRNAs, which means about 100,000 expressed genes. This number is still lower than the capacity of the unique DNA fraction in an organism. These arguments lead to the conclusion that the vast majority of cellular DNA isn't functional.

Genes that are expressed usually have **introns** that interrupt the coding sequences. A typical eukaryotic gene, therefore, consists of a set of sequences that appear in mature mRNA (called **exons**) interrupted by introns. The regions between genes are likewise not expressed, but may help with chromatin assembly, contain promoters, and so forth. See Figure 12-6.

Intron sequences contain some common features. Most introns begin with the sequence GT (GU in RNA) and end with the sequence AG. Otherwise, very little similarity exists among them. Intron sequences may be large relative to coding sequences; in some genes, over 90 percent of the sequence between the 5' and 3' ends of the mRNA is introns. RNA polymerase transcribes intron sequences. This means that eukaryotic mRNA precursors must be **processed** to remove introns as well as to add the caps at the 5' end and polyadenylic acid (poly A) sequences at the 3' end.

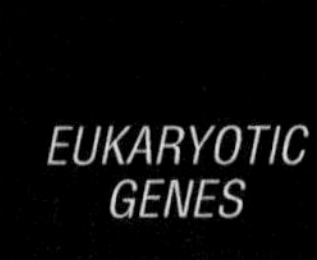

Figure 12-6

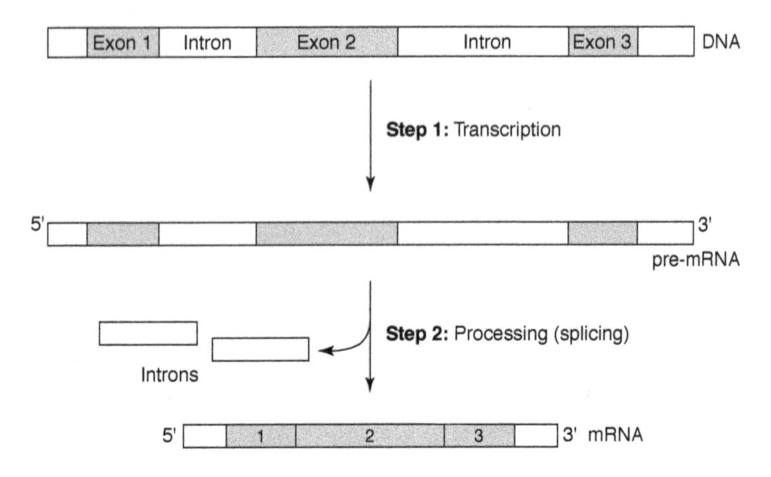

Eukaryotic genes may be clustered (for example, genes for a metabolic pathway may occur on the same region of a chromosome) but are independently controlled. Operons or polycistronic mRNAs do not exist in eukaryotes. This contrasts with prokaryotic genes, where a single control gene often acts on a whole cluster (for example, *lac*I controls the synthesis of β-galactosidase, permease, and acetylase).

One well-studied example of a clustered gene system is the mammalian globin genes. Globins are the protein components of hemoglobin. In mammals, specialized globins exist that are expressed in embryonic or fetal circulation. These have a higher oxygen affinity than adult hemoglobins and thus serve to "capture" oxygen at the placenta, moving it from the maternal circulation to that of the developing embryo or fetus. After birth, the familiar mature hemoglobin (which consists of two alpha and two beta subunits) replaces these globins. Two globin clusters exist in humans: the alpha cluster on chromosome 16, and the beta cluster on chromosome 11, as shown in Figure 12-7.

Figure 12-7

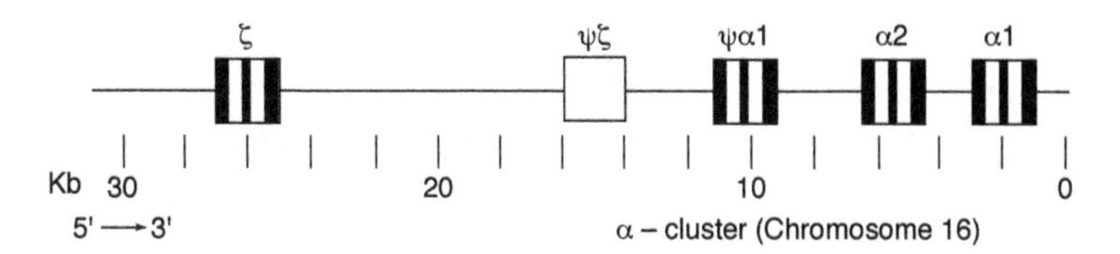

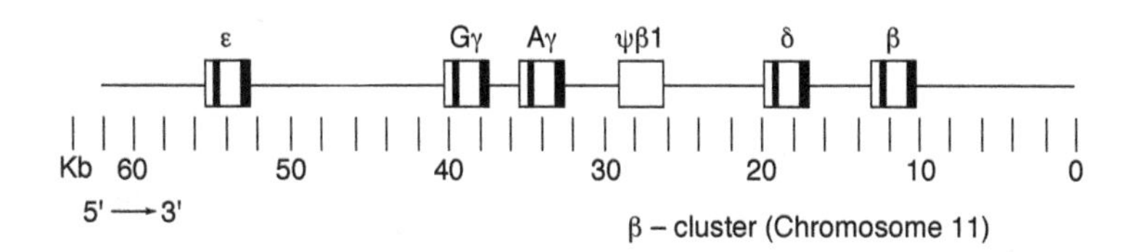

These clusters, and the gene for the related protein myoglobin (see Chapter 6 in Volume 1), probably arose by duplication of a primordial gene that encoded a single heme-containing, oxygen-binding protein. Within each cluster is a gene designated with the Greek letter Ψ. These are **pseudogenes**—DNA sequences related to a functional gene but containing one or more mutations so that it isn't expressed.

The information problem of eukaryotic gene expression therefore consists of several components: **gene recognition, gene transcription**, and **mRNA processing**. These problems have been approached biochemically by analyzing the enzyme systems involved in each step.

Transcription

Eukaryotic transcription uses three distinct RNA polymerases, which are specialized for different RNAs. **RNA polymerase I** makes Ribosomal RNAs, **RNA polymerase II** makes messenger RNAs, and **RNA polymerase III** makes small, stable RNAs such as transfer RNAs and 5S ribosomal RNA. Eukaryotic RNA polymerases are

differentiated by their sensitivity to the toxic compound, **α-amanitin**, the active compound in the poisonous mushroom *Aminita phalloides,* or "destroying angel." RNA polymerase I is not inhibited by α-amanitin, RNA polymerase II is inhibited at very low concentrations of the drug, and RNA polymerase III is inhibited at high drug concentrations.

Eukaryotic transcription is dependent on several sequence and structural features. First, actively transcribing genes have a "looser," more accessible chromatin structure. The nucleosomes are not as condensed as in other forms of chromatin, especially heterochromatin, and they often do not contain histone H1. The DNA in the promoter region at the 5' end of the gene may not be bound into nucleosomes at all. In this way, the promoter sequences are available for binding to protein **transcription factors**—proteins that bind to DNA and either repress or stimulate transcription. In addition to promoter sequences, other nucleotide sequences termed **enhancers** can affect transcription efficiency. Enhancers bind to specialized protein factors and then stimulate transcription. The difference between enhancers and factor-binding promoters depends on their site of action. Unlike promoters, which only affect sequences immediately adjacent to them, enhancers function even when they are located far away (as much as 1,000 base pairs away) from the promoter. Both enhancer-binding and promoter-binding transcription factors recognize their appropriate DNA sequences and then bind to other proteins—for example, RNA polymerase, to help initiate transcription. Because enhancers are located so far from the promoters where RNA polymerase binds, enhancer interactions involve bending the DNA to make a loop so the proteins can interact.

Ribosomal RNA synthesis

Most of the RNA made in the cell is ribosomal RNA. The large and small subunit RNAs are synthesized by RNA polymerase I. Ribosomal RNA is made in a specialized organelle, the **nucleolus**, which contains many copies of the rRNA genes, a correspondingly large number of RNA polymerase I molecules, and the cellular machinery that processes the **primary transcripts** into mature

rRNAs. RNA polymerase I is the most abundant RNA polymerase in the cell, and it synthesizes RNA at the fastest rate of any of the polymerases. The genes for rRNA are present in many copies, arranged in tandem, one after the other. Each transcript contains a copy of each of three rRNAs: the 28S and 5.8S large subunit RNAs and the small subunit 18S RNA, in that order. The rRNA promoter sequences extend much further upstream than do prokaryotic promoters. The transcription of rRNA is very efficient. This is necessary because each rRNA transcript can only make one ribosome, in contrast to the large number of proteins that can be made from a single mRNA.

The individual ribosomal RNAs must be processed from the large precursor RNA that is the product of transcription. The primary transcript contains small and large subunit RNAs in the order: 28S—5.8S—18S. Processing involves the modification of specific nucleotides in the rRNA, followed by cleavage of the transcript into the individual RNA components. See Figure 12-8.

Figure 12-8

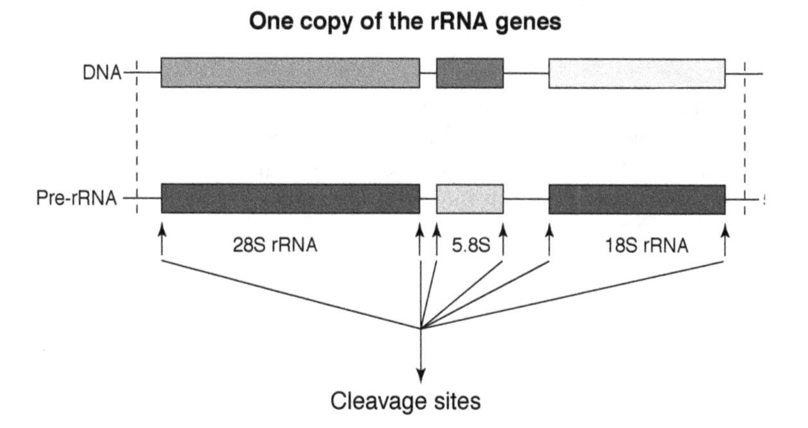

Messenger RNA transcription

RNA polymerase II transcribes messenger RNA and a few other small cellular RNAs. Class II promoters are usually defined by their sensitivity to α-amanitin. Like prokaryotic promoters, many class II promoters contain two conserved sequences, called the CAAT and TATA boxes. The TATA box is bound by a specialized transcription factor called TBP (for TATA-Binding-Factor). Binding of TBP is required for transcription, but other proteins are required to bind to the upstream (and potentially downstream) sequences that are specific to each gene. Like prokaryotic transcripts, eukaryotic RNAs are initiated with a nucleoside triphosphate. Termination of eukaryotic mRNA transcription is less well understood than is termination of prokaryotic transcription, because the 3' ends of eukaryotic mRNAs are derived by processing. See Figure 12-9.

Figure 12-9

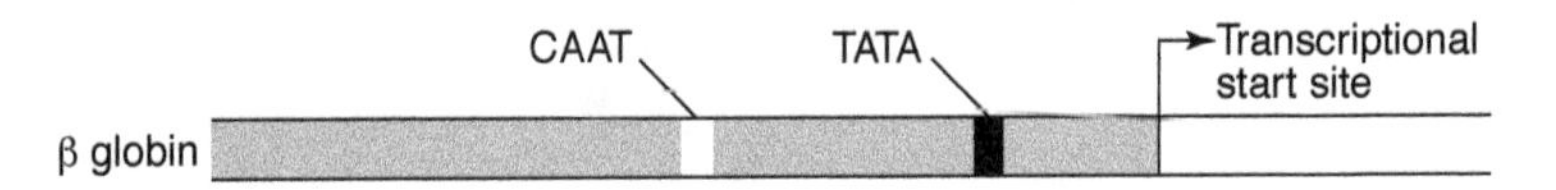

Transfer and 5S ribosomal RNA transcription

RNA polymerase III transcribes 5S rRNA and tRNA genes. The "promoter" of these transcripts can actually be located *inside the gene itself,* in contrast to all the other promoters discussed earlier. See Figure 12-10.

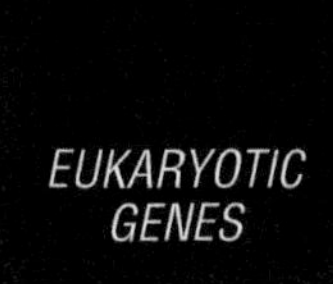

Figure 12-10

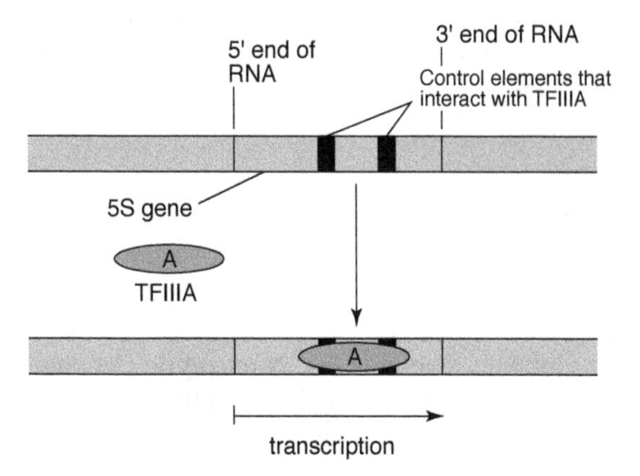

The 5' sequence is not essential for accurate transcription initiation. When the region extending from the 5' end of the gene (that is, the part that would normally be considered to be the promoter) is deleted, RNA synthesis is carried out just as efficiently as on the native gene. The new 5' end of the transcript is complementary to whatever sequences take the place of the natural ones. Furthermore, initiation is only affected when sequences *within* the 5S rRNA gene are disrupted. The molecular explanation for this phenomenon is as follows:

1. A protein factor binds to the 5S rRNA gene. Binding is at the internal sequence that is required for accurate initiation.

2. The bound factor then interacts with RNA polymerase III, which is then capable of initiation. During transcription, the multiple protein factors (called TFIIIs) remain bound to the transcribing gene.

RNA Processing

All classes of RNA transcripts must be processed into mature species. The reactions include several types: **Nucleolytic cleavage**, as in the separation of the mature rRNA species from the primary transcript of RNA polymerase I action; **Chain extension** (non-template-directed), as in the synthesis or regeneration of the common CCA sequence at the 3' end of transfer RNAs or of PolyA at the 3' end of mRNAs; and **Nucleotide modification**, for example, the synthesis of methylated nucleotides in tRNA or rRNA. These reactions are a feature of both prokaryotic and eukaryotic gene expression, and the biological consequences are diverse. For example, modified nucleotides can affect the way in which a tRNA recognizes different codons.

Messenger RNA processing reactions

Messenger RNA processing is complex, especially in eukaryotes. Prokaryotic mRNA processing is relatively unimportant in regulating gene expression. The chief function of prokaryotic mRNA seems to be to regulate stability. The terminator stem and loop stabilizes mRNA against nucleolytic degradation, and in some cases, removal of this structure destabilizes mRNA so that it is transcribed less efficiently.

Eukaryotic mRNA processing is much more complex and has many consequences for gene expression. Most obviously, many eukaryotic genes contain **introns**, which are found in the primary transcript. For the mRNA to be translated into a useful protein, some way to remove them from the transcript and still preserve the coding sequence of the mature messenger RNA must exist. Additionally, the 5' ends of eukaryotic mRNAs contain a specialized **cap** structure and a **3' polyA** "tail," and these are not encoded by the DNA template. Many of these reactions involve small nuclear ribonucleoprotein particles (snRNPs or "snurps") as essential components. Other submicroscopic structures in the nucleus are essential for processing and transport of the mature RNAs from the nucleus into the cytoplasm.

A soluble enzyme system carries out **cap addition** to the 5' end of mRNA *during the time that RNA polymerase II is still synthesizing the 3' region of the mRNA*. See Figure 12-11.

First, one phosphate of the 5'-terminal phosphate of the pre-mRNA transcript is cleaved to yield a diphosphate, which then attacks GTP, releasing inorganic pyrophosphate and making a **5'-5' phosphate bond**. The immature cap is then methylated. The G is methylated to make N^7-methyl G, while the first and (sometimes) the second nucleotides of the transcript are methylated on their 2'-hydroxyl groups. The cap performs two functions: First, it is usually required for binding the mRNA to the ribosome, and it also seems to allow the recognition of introns.

Polyadenylation occurs at the 3' end of the pre-mRNA. First, the pre-mRNA is cleaved when a specific sequence, AAUAAA, is present in the transcript. Cleavage of the pre-mRNA occurs about 20 or so nucleotides "downstream" (3') of the polyA signal sequence. RNA polymerase II continues on the template, sometimes for as long as 1,000 nucleotides before termination occurs. The RNA that is made during this transcription is simply degraded. Cleavage of the pre-mRNA, on the other hand, generates a free 3'-OH group, which is then extended by the enzyme poly(A) polymerase in a reaction that uses ATP and releases inorganic pyrophosphate (just as template-directed synthesis does). The polyA sequence is about 200-300 nucleotides long when it is first made. See Figure 12-12.

Figure 12-11

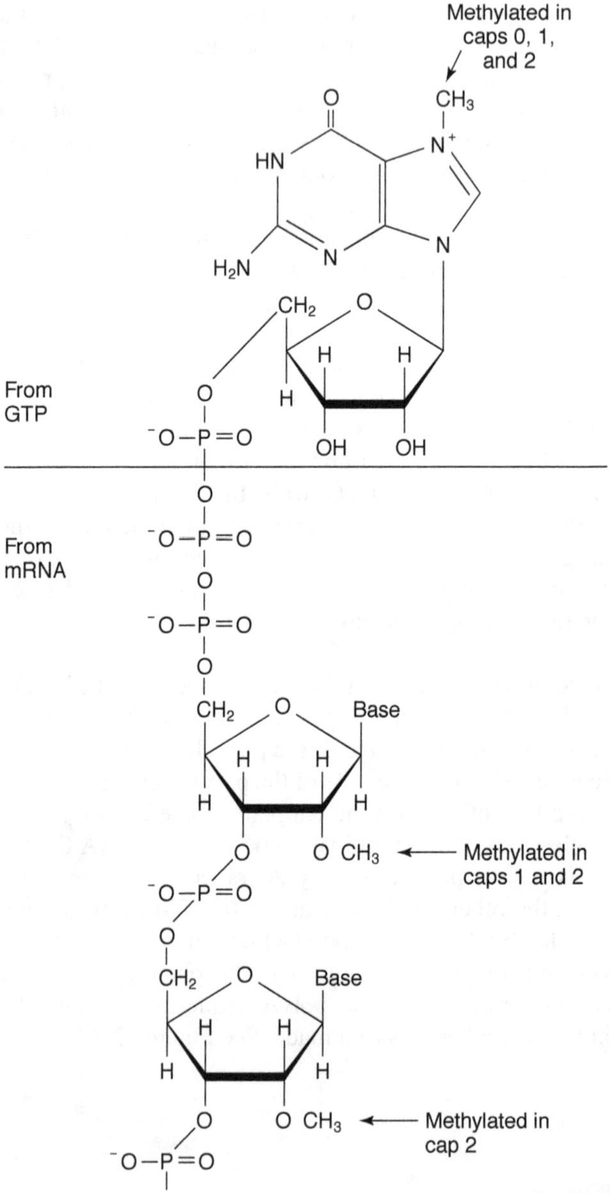

Methylated in caps 0, 1, and 2

From GTP

From mRNA

Base

Methylated in caps 1 and 2

Base

Methylated in cap 2

Figure 12-12

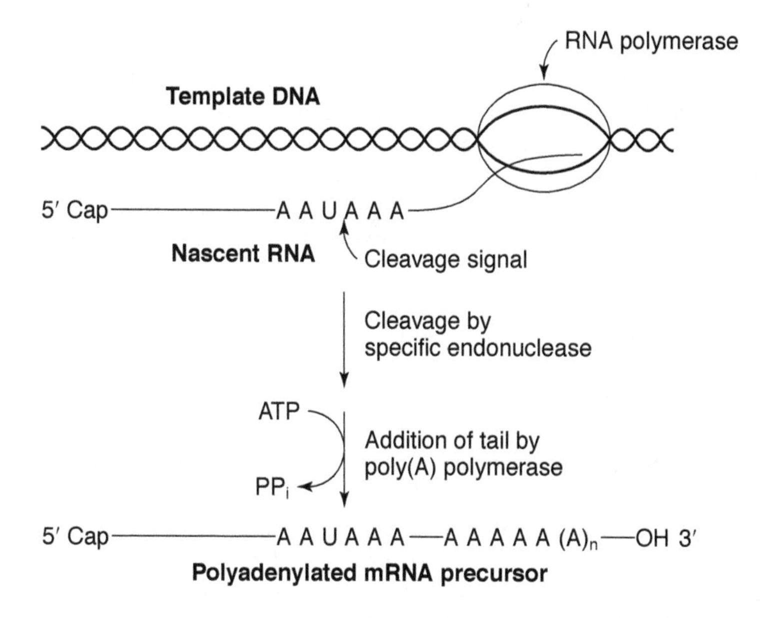

Both capping and poly A formation precede intron removal. Thus, the first steps of processing result in a pre-mRNA that has a 5' cap and a 3' polyA tail, but with all its introns present.

The term **splicing** refers to the process by which introns are removed and the mRNA put back together to form a continuous coding sequence in the 5'-3' direction. Remembering how accurate this process must be is important. If only a single nucleotide of an intron were left in the processed mRNA, the protein made from that mRNA would be non-functional, because the ribosome would read the wrong codons. The cellular machinery that splices pre-mRNAs uses information at the splice junctions to determine where to cut and where to rejoin the mRNA. Removal of introns from transcripts containing more than one intron usually occurs in a preferred but not exclusive order. Several "pathways" are used.

The small nuclear ribonucleoprotein particles (snRNPs or "snurps") that carry out the splicing reaction use **RNA-RNA base-pairing** to select the splice sites. Almost all intron-exon junctions contain the sequence AG-GU with the GU beginning the intron sequence. Furthermore, the consensus sequence for the beginning of the intron has a longer sequence complementary to the **U1 RNA**. Thus, assembly of the splicing complex, called the **spliceosome**, starts when the RNA component of the U1 snRNP base pairs with the junction between the 3' end of the exon and the 5' end of the intron. See Figure 12-13.

Introns end in the dinucleotide sequence AG. A region high in C and U residues, called the *polypyrimidine tract,* precedes this sequence. Upstream of the polypyrimidine tract is a sequence that has the (more or less) -consensus 5' AUCUAACA -3', which is partially complementary to a sequence in U2 RNA, 5' UGUAGUA 3'. One of the As in the intron sequence is bulged out when the intron-U2 junction is formed. This bulged A is important in the splicing reaction. Spliceosome assembly also involves three other snRNPs and a variety of accessory proteins, including those that recognize the polypyrimidine tract and other proteins that unwind the pre-mRNA.

The actual splicing reaction involves an unusual RNA reaction. The pre-mRNA is cleaved at the junction so that the intron begins with a 5'-phosphate. This pGU end of the intron then is joined to the free 2' OH of the bulged A at the consensus AUCUAACA sequence. The result is an unusual A with all three of its hydroxyl groups bound into a phosphodiester linkage. The 5' and 3' hydroxyls are in the linear chain of the intron, while the 2' OH forms a 2',5' phosphodiester linkage with the 5' phosphate of the intron's G. This RNA structure is called a **lariat**. The formation of the lariat determines the site of cleavage at the 3' end of the intron. Cleavage of the intron-exon junction and joining of the two exon sequences occurs at the first AG downstream of the lariat branch point.

Figure 12-13

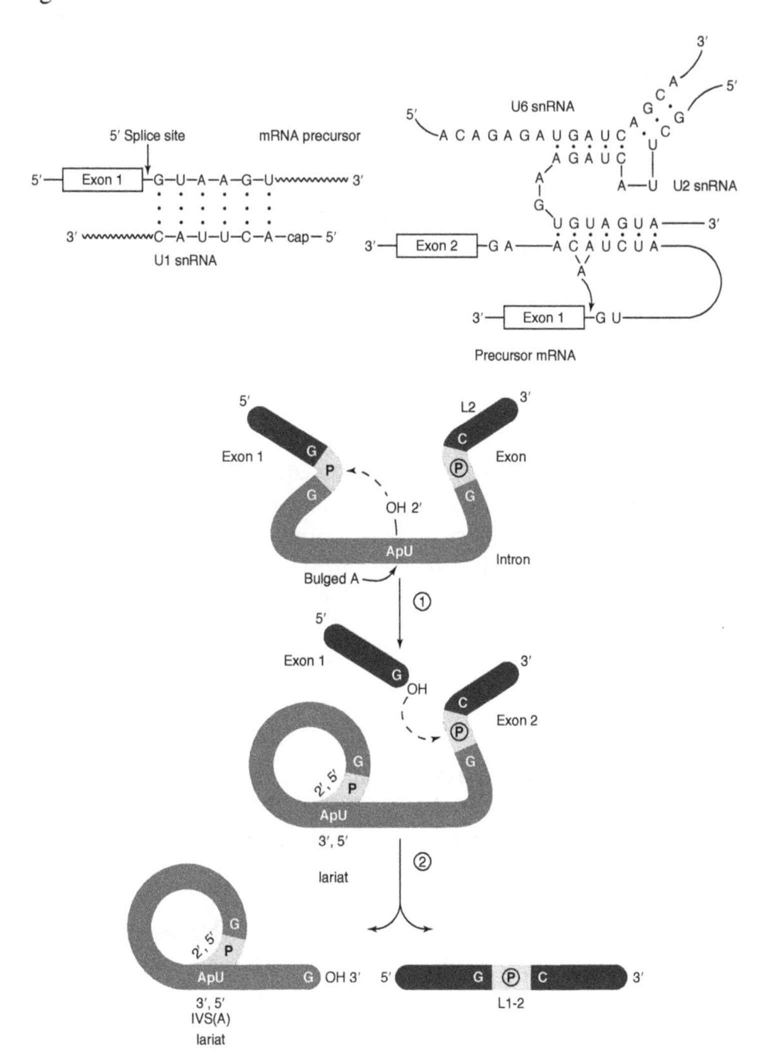
5′ Splice site
mRNA precursor
5′ Exon 1 G–U–A–A–G–U 3′
3′ C–A–U–U–C–A–cap– 5′
U1 snRNA
U6 snRNA
5′ A C A G A G A U G A U C
A G A U C
U2 snRNA
U G U A G U A 3′
3′ Exon 2 G A A C A U C U A
A
3′ Exon 1 G U
Precursor mRNA
5′
3′
L2
Exon 1
Exon
G
P
C
OH 2′
ApU
Intron
Bulged A
①
Exon 1
OH
Exon 2
2′, 5′
3′, 5′
lariat
②
OH 3′
IVS(A)
L1-2

In summary, mRNA processing requires an interconnected set of reactions: Cap and polyA synthesis, followed by intron removal. Soluble enzymes catalyze the former two reactions, while the latter set of reactions involves both RNA and protein components.

This scheme for eukaryotic mRNA processing is consistent with defects in various genetic diseases. For example, phenylketonuria (PKU) is a condition characterized by an autosomal recessive gene that results from a deficiency of the enzyme phenylalanine hydroxylase, which carries out the conversion of phenylalanine to tyrosine, which is subsequently broken down by other enzymes to TCA cycle intermediates. Children with the disease must be given a synthetic low-phenylanine diet, or mental retardation will result. The most common mutation is the conversion of an intron 5' GU sequence to an AU, which results in the synthesis of mRNA that can't be translated into active protein because intron sequences remain after processing. Similarly, some forms of the disease β-thalassemia result from defective mRNA processing reactions due to base changes in the gene for β-globin. The defective mRNA that is produced in these reactions codes for mutated β-globin that doesn't function as an oxygen carrier.

RNA catalysis

Surprisingly, RNA can catalyze biochemical reactions. Most of these "ribozymes" work on RNA substrates (often themselves) but one case is known where a naturally occurring RNA carries out a reaction not involving RNA specifically.

Thomas Cech and his coworkers discovered RNA catalysis in the early 1980's while studying the removal of an intron from the ribosomal RNA of a small ciliated protozoan, *Tetrahymena.* They established the reaction in vitro and then purified the components of the reaction mixture to isolate what they thought would be the enzyme responsible. To their surprise, they could remove all the detectable proteins from the mixture and still get intron removal to occur efficiently.

Subsequently, several other kinds of ribozymes were discovered: ribonuclease P, which cleaves transfer RNA precursors; another variety of intron from fungi, which carries out its own removal; a number of viral RNAs, which cleave themselves from large end-to-end intermediates to give genomic-sized RNAs; and, most remarkably, the ribosome. Originally, all of these RNAs were thought to have a similar mechanism; however, this may be an oversimplification, and more RNA-catalyzed reactions may exist (perhaps in the spliceosome) waiting to be characterized.

Eukaryotic Translation

The broad outlines of eukaryotic protein synthesis are the same as in prokaryotic protein synthesis. The genetic code is generally the same (some microorganisms and eukaryotic mitochondria use slightly different codons), rRNA and protein sequences are recognizably similar, and the same set of amino acids is used in all organisms. However, specific differences exist between the two types of protein synthesis at all steps of the process.

Initiation

Both prokaryotes and eukaryotes initiate protein synthesis with a specialized methionyl-tRNA in response to an AUG initiation codon. Eukaryotes, however, use an initiator met—$tRNA^{met}{}_{I}$—that is not formylated. Recognition of the initiator AUG is also different. Only one coding sequence exists per eukaryotic mRNA, and eukaryotic mRNAs are capped. Initiation, therefore, uses a specialized cap-binding initiation factor to position the mRNA on the small ribosomal subunit. Usually, the first AUG after the cap (that is, 3' to it) is used for initiation.

Elongation

Most differences in elongation result from the fact that the eukaryotic cell has different compartments, which are separated by membranes. Both prokaryotic and eukaryotic cells, of course, have an inside and outside; however, eukaryotic proteins can be targeted to, for example, the mitochondrion.

Translating ribosomes in eukaryotes are located in different places in the cell depending on the fate of their proteins. Free polysomes are in the cytoplasm and synthesize cytoplasmic proteins and those that are bound for most intracellular organelles, for example, the nucleus. Members of the second class of polysomes, membrane-bound polysomes, are attached to the endoplasmic reticulum (forming the **rough ER**), and synthesize exported proteins. In cells that are actively secreting enzymes or hormones (for example, those in the pancreas), most of the protein synthesis occurs on the rough ER.

The messages encoding exported proteins must be recognized. For example, the digestive proteases are madc in the pancreas. If they are released into the pancreas rather than the intestine, they will self-digest the organ that makes them. Carrying out this export efficiently is obviously important. The **signal hypothesis** explains how proteins destined for export are discriminated. Proteins that are destined for export contain a short (less than 30 amino acids long) sequence made up of **hydrophobic** amino acids at their amino terminus. Because peptide synthesis occurs in the amino-to-carboxy direction, the signal peptide is the first part of the protein that is made. Signal peptides are not found in most mature secreted proteins because they are cleaved from the immature proteins during the secretion and maturation process. See Figure 12-14.

The process of protein export involves a small, cytoplasmic ribonucleoprotein particle (the **Signal Recognition Particle** or **SRP**) with the signal coding mRNA sequence and/or the signal peptide itself. This interaction stops translation of the protein. Then, the stalled or arrested ribosome moves to the endoplasmic reticulum (ER). A receptor on the ER binds the SRP.

Figure 12-14

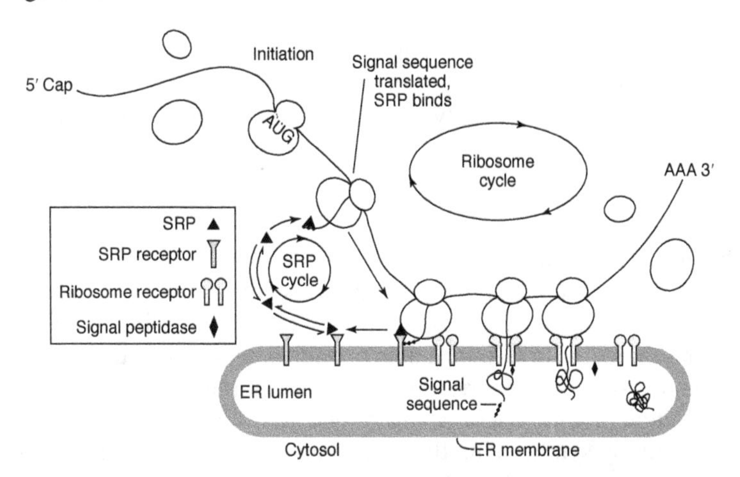

Once the complex containing ribosome, mRNA, signal peptide, and SRP is "docked" onto the membrane, SRP leaves the complex and the ribosome resumes translation. The signal peptide inserts across the membrane; this insertion is dependent on the hydrophobic nature of the signal sequence. The rest of the protein follows the signal sequence across the membrane, like thread through the eye of a needle. The protein folds into its secondary and tertiary structure in the **lumen** (inside cavity) of the ER. The signal sequence is cut away from the protein either during translation (cotranslational processing) or, less often, after the protein is released from the ribosome (post-translational processing). After the polypeptide chain is completed, the ribosome is released from the ER and is ready to initiate synthesis of a new protein. Secreted proteins are also made by prokaryotes by using a signal sequence mechanism, with the cell membrane taking the place of the ER membrane.

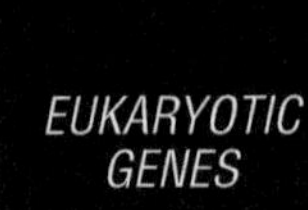

This scheme can also accommodate the synthesis of membrane-bound proteins. In this case, the protein is not released into the lumen of the ER, but rather stays bound to the membrane. One or more **anchor sequences** (or "stop-transfer sequences") in the newly made protein keeps the partially folded region of the protein in the membrane.

Protein glycosylation

Many eukaryotic membrane-bound and secreted proteins contain a complex of sugar residues bound to the side chains of either asparagine to make N-linked sugar residues or serine and threonine to make O-linked sugar residues. The core of the glycosyl complex is assembled, not by adding sugars one after another to the protein chain, but by synthesizing the core oligosaccharide on a membrane lipid, **dolichol phosphate**. See Figure 12-15.

Figure 12-15

Isoprenoid unit

$$^{-}O-\overset{O}{\overset{\|}{C}}(O^{-})-O-CH_2-CH_2-\overset{H}{\underset{CH_3}{C}}-CH_2-\left(CH_2-CH=\overset{CH_3}{C}-CH_2\right)_n-CH_2-CH=\overset{CH_3}{C}-CH_3$$

Dolichol phosphate
(n= 15-19)

Man —α1.2— Man —α1.6— Man

Man —α1.2— Man —α1.3— Man

Man —α1.6— Man —β1.4— GlcNAc —β1.4— GlcNAc—O—P(=O)(O⁻)—O—P(=O)(O⁻)—O—Dol

(Glc)₃ —— Man —α1.2— Man —α1.2— Man —α1.3— Man

Dolichol is a long chain of up to twenty 5-carbon **isoprenoid units.** The core oligosaccharide assembled on dolicol phosphate ultimately contains 14 saccharides that are transferred all at once to an asparagine residue on the protein. After the glycoprotein is assembled, the core oligosaccharide is trimmed by removal of the three glucose residues from the end, making a **high-mannose** oligosaccharide on the protein. The high-mannose oligosaccharide is moved from the ER for further modification in the **Golgi complex**, a structure composed of layered intracellular membranes. The completed glycoprotein moves to the plasma membrane in a membrane granule, which buds off from the Golgi. The granule fuses with the plasma membrane to release its contents. Binding to receptors in the plasma membrane brings some lysosomal enzymes back into the cell. The enzyme-receptor complex is taken into the cell in a process that is essentially the reverse of secretion, although it involves different membrane proteins.

Eukaryotic Transcriptional Control

Transcriptional control in eukaryotes can be accomplished at several levels. Chromatin structure can control transcription. The formation of so-called hypersensitive sites (sites where the DNA is not bound into nucleosomes) allows protein factors and RNA polymerase to access the DNA. This is necessary for transcription to occur, but hypersensitive sites are not enough. The removal of histone H1 allows transcription to occur from a chromatin domain. Some protein factors (for example, TBF) may be bound to a promoter region even if the gene is not being transcribed. TBF also is necessary but not sufficient for transcription.

Transcription control factors promote or prevent RNA polymerase binding. Various *trans*-acting factors (proteins) bind at specific *cis*-acting sequences. These factors can bind upstream of the promoter. Other factors bind to enhancer sequences and the chromatin folds to allow the enhancer-binding factors to bind to the proteins at

the promoter region or at the upstream sequences. Protein-protein interactions between bound factors contribute to transcriptional activation. Developmental gene regulation can occur through protein factors—for example, by the presence of protein at different positions in the embryo.

DNA binding proteins have common structures and means of recognizing DNA sequence. The specificity of the interaction depends on the amino acids in the protein that are available to encounter specific structures in DNA sequences. For example, the amido group on glutamine or asparagine may provide a hydrogen-bond donor to the Oxygen at position 6 of guanosine. See Figure 12-16.

Two common DNA-binding structures are found in a variety of transcriptional control proteins. The **helix-turn-helix** motif allows interaction with DNA sequences. The two α-helices are positioned at an angle to each other. One α-helix (the binding helix) contacts the major groove of the DNA molecule. The other α-helix positions the binding helix relative to the DNA. Transcriptional control proteins can have other domains that allow their interaction with other transcription factors; these protein-protein interactions allow multiple binding events to occur. Helix-turn-helix proteins are found in both prokaryotic and eukaryotic systems. See Figure 12-17.

Figure 12-16

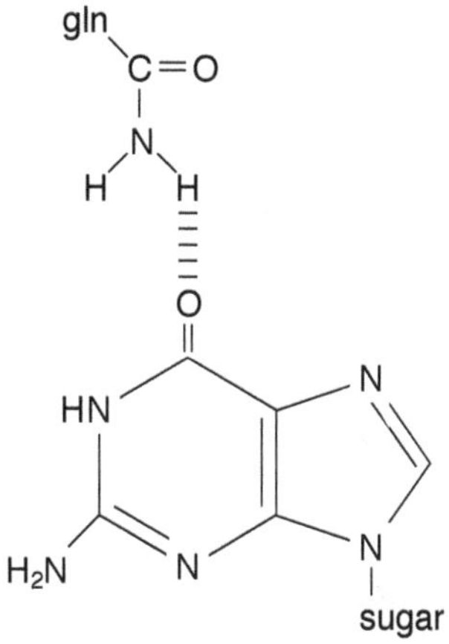

Figure 12-17

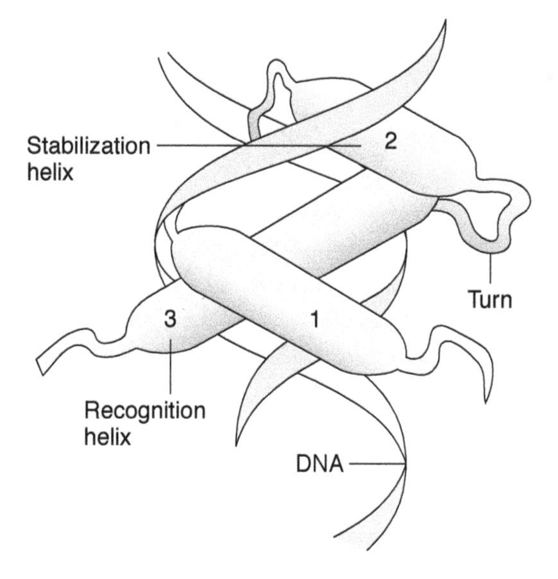

Helix-Turn-Helix

Zinc-fingers are common in DNA-binding proteins of eukaryotes but are not found in prokaryotes. Examples of zinc-finger proteins include the RNA polymerase III transcription factor TFIIIA, steroid receptors, and some gene products that control development. The zinc-finger consists of pairs of cysteine and/or histidine residues within an α-helix. These residues bind tightly to a Zn^{2+} ion, which allows the α-helical amino acids to interact with specific sequences. See Figure 12-18.

Figure 12-18

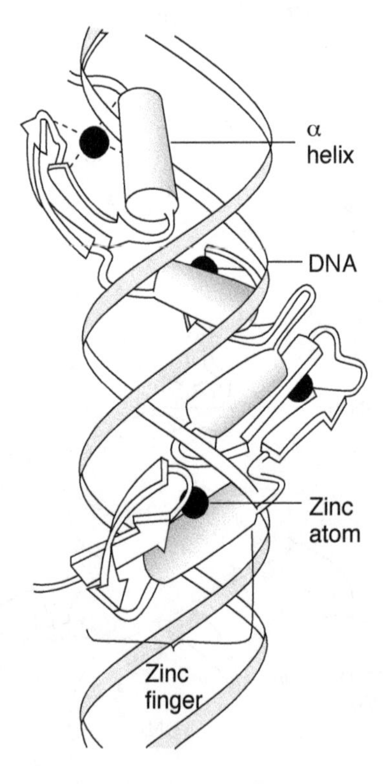

Zinc Finger

Again, protein-protein contacts allow for specific interaction between different proteins. These contacts are hypothesized to occur by the interaction of hydrophobic domains sometimes called "leucine zippers" to denote their amino acid composition as well as their function.

Translational Control

The control of gene expression can also utilize translational mechanisms. These mechanisms are usually directed at initiation. For example, in response to virus challenge, the protein **interferon** is released and turns off protein synthesis in neighboring cells by a dual mechanism. First, interferon induces mRNA degradation. In response to interferon, an unusual RNA-related molecule, **2',5'-oligoadenylic acid** is made. This stimulates a cellular endonuclease, which degrades cellular RNA to inhibit cell growth. Thus, interferon is a mechanism to limit the damage from a virus infection. A double-stranded RNA cofactor is required for 2',5'-ligoA synthesis. Cells don't usually have double-stranded RNA, but RNA viruses require double-stranded RNA for replication. This limits interferon-induced cell damage largely to virus-infected cells. Interferon also induces phosphorylation of eIF_2,which further inhibits initiation.

A similar mechanism operates in the control of globin mRNA translation. The production of heme and of globin must be closely coordinated, because hemoglobin contains exactly one heme and two each of the α- and β-globins. The major mRNAs in the developing red blood cell are those for the globin proteins. This means that changes in the activity of the ribosomes will affect globin production, primarily.

The translation of globin mRNA is dependent on free heme. The absence of heme leads to phosphorylation of eIF_2 through a **hemin-controlled inhibitor**. In the absence of heme, the hemin-controlled inhibitor protein is an active protein kinase, which phosphorylates

eIF_2. When heme is again available, the hemin-controlled inhibitor binds heme and is inactivated, allowing protein synthesis to resume. See Figure 12-19.

Figure 12-19

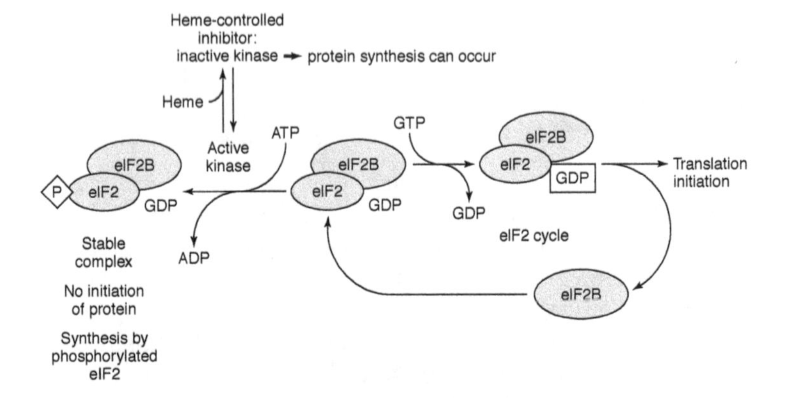

Notes

Notes

Notes

Notes

Printed in the USA
CPSIA information can be obtained
at www.ICGtesting.com
LVHW020954040124
768142LV00011B/341

9 780764 585623